BiCMOS TECHNOLOGY AND APPLICATIONS

THE KLUWER INTERNATIONAL SERIES IN ENGINEERING AND COMPUTER SCIENCE

VLSI, COMPUTER ARCHITECTURE AND DIGITAL SIGNAL PROCESSING

Consulting Editor

Jonathan Allen

Other books in the series:

Logic Minimization Algorithms for VLSI Synthesis. R.K. Brayton, G.D. Hachtel, C.T. McMullen, and Alberto Sangiovanni-Vincentelli. ISBN 0–89838–164–9.
Adaptive Filters: Structures, Algorithms, and Applications. M.L. Honig and D.G. Messerschmitt. ISBN 0–89838–163–0.
Introduction to VLSI Silicon Devices: Physics, Technology and Characterization. B. El-Kareh and R.J. Bombard. ISBN 0–89838–210–6.
Latchup in CMOS Technology: The Problem and Its Cure. R.R. Troutman. ISBN 0–89838–215–7.
Digital CMOS Circuit Design. M. Annaratone. ISBN 0–89838–224–6.
The Bounding Approach to VLSI Circuit Simulation. C.A. Zukowski. ISBN 0–89838–176–2.
Multi-Level Simulation for VLSI Design. D.D. Hill and D.R. Coelho. ISBN 0–89838–184–3.
Relaxation Techniques for the Simulation of VLSI Circuits. J. White and A. Sangiovanni-Vincentelli. ISBN 0–89838–186–X.
VLSI CAD Tools and Applications. W. Fichtner and M. Morf, Editors. ISBN 0–89838–193–2.
A VLSI Architecture for Concurrent Data Structures. W.J. Dally. ISBN 0–89838–235–1.
Yield Simulation for Integrated Circuits. D.M.H. Walker. ISBN 0–89838–244–0.
VLSI Specification, Verification and Synthesis. G. Birtwistle and P.A. Subrahmanyam. ISBN 0–89838–246–7.
Fundamentals of Computer-Aided Circuit Simulation. W.J. McCalla. ISBN 0–89838–248–3.
Serial Data Computation. S.G. Smith and P.B. Denyer. ISBN 0–89838–253–X.
Phonologic Parsing in Speech Recognition. K.W. Church. ISBN 0–89838–250–5.
Simulated Annealing for VLSI Design. D.F. Wong, H.W. Leong, and C.L. Liu. ISBN 0–89838–256–4.
Polycrystalline Silicon for Integrated Circuit Applications. T. Kamins. ISBN 0–89838–259–9.
FET Modeling for Circuit Simulation. D. Divekar. ISBN 0–89838–264–5.
VLSI Placement and Global Routing Using Simulated Annealing. C. Sechen. ISBN 0–89838–281–5.
Adaptive Filters and Equalisers. B. Mulgrew, C.F.N. Cowan. ISBN 0–89838–285–8.
Computer-Aided Design and VLSI Device Development, Second Edition. K.M. Cham, S-Y. Oh, J.L. Moll, K. Lee,
   P. Vande Voorde, D. Chin. ISBN: 0–89838–277–7.
Automatic Speech Recognition. K-F. Lee. ISBN 0–89838–296–3.
Speech Time-Frequency Representations. M.D. Riley. ISBN 0–89838–298–X
A Systolic Array Optimizing Compiler. M.S. Lam. ISBN: 0–89838–300–5.
Algorithms and Techniques for VLSI Layout Synthesis. D. Hill, D. Shugard, J. Fishburn, K. Keutzer. ISBN: 0–89838–301–3.
Switch-Level Timing Simulation of MOS VLSI Circuits. V.B. Rao, D.V. Overhauser, T.N. Trick, I.N. Hajj. ISBN 0–89838–302–1
VLSI for Artificial Intelligence. J.G. Delgado-Frias, W.R. Moore (Editors). ISBN 0–7923–9000–8.
Wafer Level Integrated Systems: Implementation Issues. S.K. Tewksbury. ISBN 0–7923–9006–7
The Annealing Algorithm. R.H.J.M. Otten & L.P.P.P. van Ginneken. ISBN 0–7923–9022–9.
VHDL: Hardware Description and Design. R. Lipsett, C. Schaefer and C. Ussery. ISBN 0–7923–9030–X.
The VHDL Handbook. Dr. Coelho. ISBN 0–7923–9031–8.
Unified Methods for VLSI Simulation and Test Generation. K.T. Cheng and V.D. Agrawal. ISBN 0–7923–9025–3
ASIC System Design with VHDL: A Paradigm. S.S. Leung and M.A. Shanblatt. ISBN 0–7923–9032–6.

BiCMOS TECHNOLOGY AND APPLICATIONS

edited by
A.R. Alvarez
Aspen Semiconductor Corporation
(A Cypress Semiconductor Company)

Kluwer Academic Publishers
Boston/Dordrecht/London

Distributors for North America:
Kluwer Academic Publishers
101 Philip Drive
Assinippi Park
Norwell, MA 02061, USA

Distributors for all other countries:
Kluwer Academic Publishers Group
Distribution Centre
Post Office Box 322
3300 AH Dordrecht, THE NETHERLANDS

Library of Congress Cataloging-in-Publication Data

BiCMOS technology and applications.

(The Kluwer international series in engineering
and computer science ; #76. VLSI, computer architecture,
and digital signal processing)
   Includes bibliographies and index.
   1. Metal oxide semiconductors, Complementary.
2. Bipolar integrated circuits.   I. Alvarez, A. R.
(Antonio R.)   II. Series: Kluwer international
series in engineering and computer science ;
SECS #76.   III. Series: Kluwer international series
in engineering and computer science. VLSI, computer
architecture, and digital signal processing.
TK7871.99.M44B53   1989   621.381 '5   89-15310
ISBN 0-7923-9033-4

Printed in the United States of America.

Contents

Chapter 1. Introduction to BiCMOS
A.R. Alvarez (Aspen Semiconductor Corp.)

Chapter 2. Device Design
J. Teplik (Motorola Inc.)

Chapter 3. BiCMOS Process Technology
R.A. Haken, R.H. Havemann, R.H. Eklund, L.N. Hutter
(Texas Instruments, Inc.)

Chapter 4. Process Reliability

R. Lahri, S.P. Joshi, B. Bastani
(National Semiconductor Corporation)

Chapter 5. Digital Design
K. Deirling, (Dallas Semiconductor)

Chapter 6. BiCMOS Standard Memories
H.V. Tran, P.K. Fung, D.B. Scott, A.H. Shah
(Texas Instruments, Inc.)

Chapter 7. Specialty Memories
C. Hochstedler (National Semiconductor Corporation)

Chapter 8. Analog Design
H.S. Lee (Massachusetts Institute of Technology)

Foreword

The topic of bipolar compatible CMOS (BiCMOS) is a fascinating one and of ever-growing practical importance. The "technology pendulum" has swung from the two extremes of preeminence of bipolar in the 1950s and 60s to the apparent endless horizons for VLSI NMOS technology during the 1970s and 80s. Yet starting in the 1980s several limits were clouding the horizon for pure NMOS technology. CMOS reemerged as a viable high density, high performance technology. Similarly by the mid 1980s scaled bipolar devices had not only demonstrated new high speed records, but early versions of mixed bipolar/CMOS technology were being produced. Hence the paradigm of either high density or high speed was metamorphasizing into an opportunity for both speed and density via a BiCMOS approach. Now as we approach the 1990s there have been a number of practical demonstrations of BiCMOS both for memory and logic applications and I expect the trend to escalate over the next decade.

This book makes a timely contribution to the field of BiCMOS technology and circuit development. The evolution is now indeed rapid so that it is difficult to make such a book exhaustive of current developments. Probably equally difficult is the fact that the new technology opens a range of novel circuit opportunities that are as yet only formative in their development. Given these obstacles it is a herculean task to try to assemble a book on BiCMOS. Nonetheless, I am pleased to preview this volume and discuss its contents from two specific perspectives. First, the contributors provide the quintessential element ---- lots of experience. Second, the contents itself is both tutorial and state-of- the-art; a difficult gap to bridge in a single volume. In the following I will briefly discuss each point.

Putting together an edited book of chapters prepared by different authors is a difficult task. In that regard Tony Alvarez has done a yeoman's job by means of both deep knowledge in the field and an evangelist's drive and enthusiasm. Based on his experience and professional connections in the community, Tony has assembled an all-star cast of technical contributors. Certainly the cumulative experience from the Motorola, National and Texas Instruments groups provides a core of know-how and critical judgment. Moreover, the selection of other specialists from industry and academia has rounded the field of contributors to make the volume reflect a community effort rather than a parochial account of specific groups. The value of experience in this endeavor is high since BiCMOS requires a clear perspective on both technologies individually. In the case of the bipolar side this is especially important since the technology is complex and without suitable baselines the merged technology may not achieve noteworthy performance. The authors indeed have the necessary experience to provide an insightful illumination of the promising developments as well as points for future consideration.

Turning to the contents, this book provides the right balance of state-of-the-art and tutorial material. Without such a balance the effort might not clarify the juxtaposition of research frontiers versus the long history of I.C. development. In the earlier chapters of the book both methodology of technology development as well as specifics are discussed. The growing need for effective use of computer aids is clearly developed and novel methods and results are demonstrated. As for the technology itself there is a very clear picture painted which includes both the necessary device physics and technology discussion as well as the key issues of manufacturing and reliability. Especially in the area of high performance, the issues of cost and manufacturability will continue to grow in importance.

The circuit aspects of the book builds on the underlying device and technology sections. As pointed out above, the problems of innovation are just now unfolding. The experiences to date in SRAM, logic and analog circuits provide an ample set of examples to consider the future trends. While the conclusions in these areas are more tentative, the material presented and the organization sets the stage for a thoughtful evaluation of options for the future. Namely, the problems of individual circuit approaches (i.e. only MOS or bipolar) become clearer as one seeks to merge and overcome these limits. In particular the fundamental limits become clear vis-a-vis those problems resulting from technology limits. The work presented here shows a good balance of both building/testing as well as designing/calculating the options.

The crystal ball gives no clear picture of technologies that are two generations down stream. Recent panel discussions on the future of BiCMOS showed no single-minded view. Some feel that continued CMOS scaling will over take BiCMOS and erode the performance edge. On the other hand, the increasing use of differential and analog techniques in digital circuits seems clear and in that regard the control of threshold and reference voltages seems rock solid on the bipolar side. The issues of process control and system level statistics are topics which will require further evaluation ----- this obviously includes voltage supply and temperature range issues as well. Especially in that regard, the chapter of this book on Process Reliability sets the stage for critical evaluation. However, the circuit breakthroughs now possible with merged bipolar/MOS technology on a single wafer represents an opportunity as intriguing as the development of the monolithic IC itself. That is, innovations yet to come based on combined device effects pose great opportunity for innovation and change. This combined with the growing need for interface and sensor circuits leads me to conclude that BiCMOS is more than a limited window of opportunity in the evolution of independent MOS and bipolar worlds.

R.W. Dutton
Stanford University

Preface

The goal of this book is to twofold. First it attempts to provide a synthesis of available knowledge about an emerging technology ----- the combination of bipolar and MOS transistors in a common integrated circuit ---- BiCMOS. While there is a great deal of published information on the subject, it has not been compiled and digested so as to be readily available. Second, it is the intent to provide anyone with a knowledge of either CMOS or Bipolar technology/ design a reference with which they can use to make educated decisions regarding the viability of BiCMOS in their application. Given the rapid evolution in this area, this monograph cannot hope to be complete. It does however attempt to cover key basic issues in technology, process, reliability, and digital/analog design. Given the early stage of the technology, insufficient coverage could be given to computer-aided tools for circuit design. This is an area that is ripe for research and development and will hopefully be included in any future edition. By paying attention to basics it is hoped that even in the fast paced field of integrated electronics, obsolescence will remain at bay for at least several years.

This book should be of interest to practicing integrated circuit engineers as well as technical managers trying to evaluate business issues related to BiCMOS. While not intended as a textbook, this book can be used in at the graduate level for a special topics course in BiCMOS. A general knowledge in device physics, processing and circuit design is assumed. Given the division of the book, it lends itself well to a two part course; one on technology and one on design. This will hopefully provide advanced students a good understanding of tradeoffs between bipolar and MOS devices and circuits.

The material covered in this book is divided into eight chapters. The first half of the book deals primarily with process and technology issues while the second half covers applications. In Chapter I BiCMOS is introduced and an overview of both technology and process issues are reviewed. A first-order cost model is developed to allow weighing of the cost-benefit of the technology. The chapter concludes with an overview of applications ranging from memories to analog design. The next three chapters (Chapters 2 - 4) thoroughly cover the device, process and reliability aspects of BiCMOS. In Chapter 2 a framework for device design is developed that is equally applicable to CMOS, Bipolar or BiCMOS technology. A response-surface based statistical methodology is presented as one approach to handling the complexity of VLSI device/technology design. Process integration issues are handled in Chapter 3 for both digital and analog applications. The wealth of material covered in this chapter ranges from process manufacturability to the intricacies of the individual process steps that comprise advanced BiCMOS technologies. A framework for process reliability assurance is covered in detail in Chapter 4. The systematic approach toward designing in reliability at the component level from the beginning is a critical aspect to the acceptance of a new technology. Issues covered range from hot carriers in both NPN and MOS transistors to ESD and electromigration.

Having covered the basic device and process issues, the second half of the book is dedicated to applications. Chapter 5 lays the foundation for digital design of BiCMOS circuits. Basic BiCMOS gates and building blocks are explained in detail. Speed and power issues are discussed to aid designers in deciding in mixing CMOS, bipolar, and BiCMOS logic. It finishes with an overview of ASIC circuits. Memories is one area in which BiCMOS has found ready acceptance. In Chapter 6 the building blocks required to design a high performance static or dynamic RAM are analyzed. Various approaches to BiCMOS RAM design are reviewed and implementation tradeoffs are elucidated. Specialty memories is evolving area for application of "smart" or application specific memories. Given the potential versatility BiCMOS brings to specialty memories it was felt that a chapter (Chapter 7) dedicated to this topic was in order. This chapter is unique that besides demonstrating the utility of BiCMOS in a number of different applications, it offers one of the most lucid and complete accounts of this rapidly growing field. Finally, in Chapter 8 analog aspects of BiCMOS are covered. This chapter is more tutorial in nature given the potential breath of applications possible. Starting from the basic analog cell, Chapter 8 provides a clear explanation of BiCMOS vs CMOS tradeoffs enabling analog design engineers to make educated tradeoffs when designing in BiCMOS.

When first approached by Carl Harris to prepare a manuscript on BiCMOS I quickly told him that it would be too large a task for any one person or even one company at this time. Carl however was persistent and encouraged me to entertain the idea of an edited book. Given the popularity of the BiCMOS short course sponsored by the IEEE at the 1987 International Electron Device Meeting (coordinated by Roger Haken, one of the book's authors) and the growing importance of BiCMOS it seemed that the job was one worth pursuing. A large number of individuals contributed to the completion of this book. I am immensely grateful to the authors for their dedicated effort. Given the newness of the field all the authors are either heavily involved in taking BiCMOS technology from the research lab to the marketplace or in expanding the understanding of the technology. Consequently to take time from their already over-burdened schedule is a clear demonstration of their dedication to the field. In order to make an edited book successful, the various chapters have to both complement each other and provide a natural progression, each building on the other. I was fortunate that from the beginning that there was an excellent spirit of cooperation among the authors. My only regret is that time pressures prevented me from providing this book an international flavor. Given the outstanding contributions made at Hitachi, Toshiba, and other companies outside the U.S. this can be viewed as a shortcoming. I do however believe that their work is adequately represented.

Not all those that contributed to this book are listed among the authors. Besides the colleagues that provided "critical" support, I am especially grateful to all the family members that had to bear the brunt of lost weekends and weeknights while the authors labored over their chapters. I personally would like to acknowledge the support of previous colleagues at Motorola (J. Saltich, K. Drozdoweicz, S. Mastroianni, and R. Roop) as well as current and previous co-workers: P. Meller, S. Arney, B. Tien, H.B. Laing, T. Hulseweh, J. Kirchgessner, A. Ballenberger, S. Cosentino, F. Ormerod, D. Schucker, R. Shrivastava, J. Arreola, M. Hartranft, N. Ratnakumar, S.Y. Pai, F. Chien, J. Stinehelfer, G. Gibbs, R. Ramirez, and K. Kanegawa. Consultation with C. Sodine of M.I.T, R. Dutton of Stanford University, and K. Mijata and T. Nagao of Hitachi provided many useful insights into the applicability of BiCMOS. Finally I would like to thank the respective companies and university for allowing participation of the authors in this book. I am especially indebted to B. Lutz at Aspen and T.J. Rodgers at Cypress both of whom supported my effort on this book in spite of the fact that they probably really felt that I should have been doing some real work.

A. R. Alvarez

BiCMOS TECHNOLOGY AND APPLICATIONS

Chapter 1

Introduction To BiCMOS
A.R. Alvarez (Aspen Semiconductor Corp.)

1.0 Introduction

BiCMOS technology combines **Bipolar** and **CMOS** transistors in a single integrated circuit. By retaining the benefits of Bipolar and CMOS, BiCMOS is able to achieve VLSI circuits with speed-power-density performance previously unattainable with either technology individually. CMOS technology maintains an advantage over Bipolar in power dissipation, noise margins, packing density, and the ability to integrate large complex functions with high yields. Bipolar technology has advantages over CMOS in switching speed, current drive per unit area, noise performance, analog capability, and I/O speed. This last point is especially significant given the growing importance of ECL I/O, historically the exclusive domain of Bipolar technology, for high speed systems [1.1]. It follows that BiCMOS technology offers the advantages of: 1) improved speed over CMOS, 2) lower power dissipation than Bipolar (which simplifies packaging and board requirements), 3) flexible I/Os (TTL, CMOS, or ECL), 4) high performance analog, and 5) latchup immunity [1.2]. Compared to CMOS, the reduced dependence on capacitive load and process/temperature variations, and the multiple circuit configurations and I/Os possible with BiCMOS greatly enhance design flexibility and can lead to reduced design cycle time. The inherent robustness of BiCMOS with respect to temperature and process variations also reduces the variability of final electrical parameters resulting in a higher percentage of prime units, an important economic consideration.

The main drawbacks to BiCMOS are higher costs and longer fabrication cycle time. Both are due to added process complexity, which will become less of a differentiator as CMOS technology complexity continues to increase. CAD compatibility, currently lagging CMOS, will improve as the technology matures and the market expands. As gate density and pin count increase, switching noise and clock skew can limit performance in TTL and CMOS compatible high speed systems. BiCMOS's ECL I/O capability, high speed, and low static power dissipation together make it an obvious choice for implementing high performance systems. But, even if BiCMOS meets all the technological criteria, market requirements must also be satisfied. Therefore a key aspect of BiCMOS's success will be the price-performance factors differentiating it from either Bipolar or CMOS.

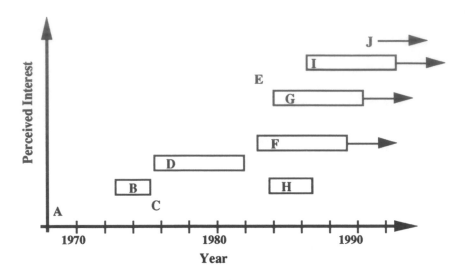

Fig. 1.1 Historical perspective of BiCMOS technology - A) 1st Publication [1.3], B) RCA Op Amps, C) Stanford DMOS, D) TI Display Drivers, E) Hitachi Memories, F) Motorola, GE Smart Power,G) Hitachi, Motorola, NEC Gate Arrays, H) SGS, Motorola High Voltage, I) National, Hitachi Fujitsu, TI Memories, J) μPs.

This chapter provides an overview and basic introduction to BiCMOS technology, process, and applications. These topics are covered in more detail in subsequent chapters. A historical overview of the evolution of BiCMOS technology from metal-gate op amps to current VLSI applications is provided in Section 1.1. Section 1.2 covers device technology aspects of BiCMOS, including a first order discussion of the basic BiCMOS inverter. This discussion illustrates some of the basic benefits of BiCMOS. Process and cost issues are covered in Section 1.3 and 1.4. Finally, a market perspective and basic applications of BiCMOS are discussed in Section 1.5. This chapter concludes with a projection of future BiCMOS performance and applications.

1.1 BiCMOS Historical Perspective

A historical perspective of BiCMOS technology is illustrated in fig. 1.1. Attempts to combine Bipolar and MOS transistors on a common integrated circuit date to 1969 [1.3]. RCA was an early leader, introducing metal-gate BiCMOS operational amplifiers in the mid 1970s [1.4]. The next major trend was high voltage BiCMOS. Pioneered at Stanford [1.5] and commercialized by Texas Instruments [1.6], this 'BIDFET^{TM}' technology, combined CMOS, Bipolar, and high voltage lateral DMOS transistors. BIDFET technology was utilized for display drivers and voltage regulators [1.7]. Most of these initial applications were analog oriented and BiCMOS technology continues to be applied in the analog and power arena [1.8-1.12]. "Smart" power applications have evolved to include extremely high current (>20 A) and voltage (>500 V) levels. A third wave of BiCMOS applications became evident in the mid 1980s.

Motivated by the power dissipation constraints of Bipolar circuits, speed limitations of MOS transistors, and the requirement for higher I/O throughput, Hitachi, Toshiba, and Motorola developed 5V digital BiCMOS technologies [1.1,1.13-1.14]. Even with compromised NPN performance, these initial offerings significantly enhanced system speed and/or reduced power. Subsequent BiCMOS technologies succeeded in eliminating performance compromises between the Bipolar and MOS transistors, resulting in extremely high performance memories, gate arrays and microprocessors [1.15-1.18]. Product growth in this high speed digital area continues to be explosive.

The evolution of high performance digital MOSFET technology is further illustrated in fig. 1.2 [1.19]. In the 1970s NMOS technology was dominant. Initially it offered high performance, low power, and process simplicity. As designs evolved additional features were demanded driving up process complexity. Larger chips led to intolerable power dissipation. Even so, NMOS advocates claimed that CMOS technology was too slow, too complex, and too area intensive to be competitive and that it would remain a niche technology. In spite of these "penalties" the late 1970s saw a gradual switch to CMOS technology. Two factors motivated the switch. Because additional process complexity had been introduced to make NMOS competitive, by 1978 the process complexity penalty incurred by switching to CMOS was negligible. This penalty was more than made up by the reduction in power dissipation. As design techniques evolved it was demonstrated that CMOS was not slower than NMOS, and that through proper chip partitioning CMOS did not require a significantly greater

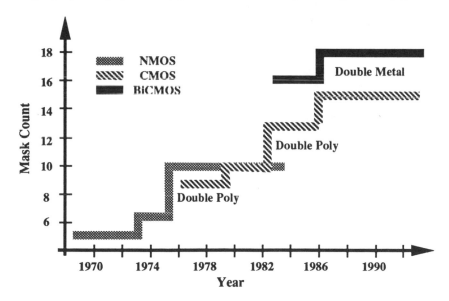

Fig. 1.2 Evolution of high performance digital MOSFET technology from NMOS to CMOS to BiCMOS. Each step in mask count is symbolic of additional complexity; ie.e depletion loads, intrinsic polysilicon load resistors, double layer metallization.

area than NMOS. In the mid 1980s, as greater performance was required, BiCMOS was seen as an alternative to CMOS in certain applications. As CMOS complexity increased, the percentage difference between CMOS and BiCMOS mask steps decreased. Therefore just as power dissipation constraints motivated the switch from NMOS to CMOS technology in the late 70s, performance requirements are motivating a switch from CMOS to BiCMOS technology in the late 80s.

In the future, both digital and digital-analog systems will exploit even more advanced BiCMOS technologies [1.20]. High speed (>60MHz) microprocessors, 10ns 1M Static RAMs, and 100K gate arrays with gate delays of 250ps or less will be available in 1990. An exciting area that is just beginning to open up is that of telecommunications [1.19]. The potential for future BiCMOS VLSI applications is covered in Section 1.5 and subsequent chapters. An excellent perspective on future BiCMOS technology and applications is found in [1.21].

1.2 BiCMOS Device Technology

BiCMOS integration presents unique challenges to the device technologist. Impurity profiles for NPNs and CMOS transistors must be optimized while simultaneously maintaining process simplicity. It follows that a rigorous device design methodology coupled with an optimized process architecture must be devised (Chapter 2). The path leading to an 'optimal' set of device parameters is circuit performance driven [1.19]. Three key circuit performance parameters that require optimization in any application are speed, power dissipation, and noise margins. The dependence of these circuit parameters on device performance and the circuit architecture determine the device parameters to be optimized, the ratio of FETs to NPNs, and consequently the process flow. For FETs speed depends on device parameters such as saturation current (Idsat), and intrinsic (Cint) and extrinsic (Cext) capacitances. These parameters in turn depend on oxide thickness, channel length and bulk doping. In FET circuits scaling the oxide thickness does not increase speed for circuits dominated by Cint as both Idsat and Cint increase at approximately the same rate. For Cext dominated circuits, scaling gate oxide or channel length will improve circuit speed. Peak NPN speed is less dependent on extrinsic capacitances. The parameters Ft, Jk and Rb, which determine circuit speed performance, are related primarily to the design of the intrinsic NPN. These device parameters depend on process parameters such as basewidth, epitaxial layer profile, emitter width and extrinsic base formation. Transistor geometry and structural makeup determine Cexts and resistances. The optimal combination of BiCMOS device parameters then depend on the circuit techniques used. If the NPNs are restricted mostly to the I/Os, relatively low performance NPNs may be satisfactory. This implies that the NPNs will present a high capacitive load and consequently thin oxide/short channel FETs are desired. If on the other hand, high performance self-aligned NPNs are used extensively in the design it is possible to relax the requirements on FET gate oxide thickness as long as short-channel effects remain under control. To determine the required FET process parameters for a given NPN design, gate

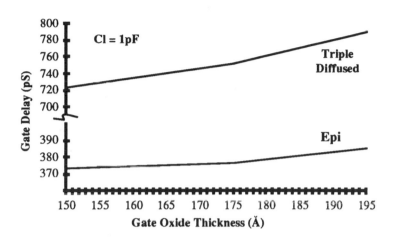

Fig. 1.3 BiCMOS Inverter gate delay as a function of gate oxide thickness for Epi and Triple Diffused NPNs; Lp = 1μm. Temperature = 25°C, Vcc = 5.0V.

delay can be studied as a function of channel length and oxide thickness (fig. 1.3). Similarly NPN Ft can be studied as a function of FET parameters to determine the required level of NPN performance (Chapter 2).

Power dissipation and noise margins in BiCMOS circuits are primarily a function of the FET threshold voltages and NPN collector and emitter resistance. Since a BiCMOS gate does not swing rail to rail, some finite power is dissipated when driving a CMOS or BiCMOS gate. This component is not significant if power dissipation is AC power (fCV^2) constrained. If necessary this leakage component of power dissipation can be reduced by increasing MOS Vts and reducing NPN resistances. Higher thresholds also improve noise margins. Unfortunately, higher thresholds result in a speed degradation. It follows that the device and circuit designer must work closely to arrive at an optimal choice of device parameters. Compared to CMOS, a BiCMOS buffer dissipates less power because of: 1) the lower signal swing, and 2) the reduction in time spent in the transition region between the gate's "on" and "off" states [1.1].

Equal area CMOS and BiCMOS inverter gate delay versus capacitive load are shown in fig. 1.4 for a 1μm process. The capacitive load at which the BiCMOS performance is equal to the CMOS (cross-over point) is critical. How this point changes as a function of device parameters such as Cj, Gm, Leff and Weff, NPN Ft, and external conditions such as temperature is one of the factors in determining the cost-effectiveness of BiCMOS in digital systems. At 25°C, typical process conditions and 5V this cross-over occurs at approximately a fanout of 1.5 - 2.0 (100 - 300fF) for simple functions like inverters and NANDs, and between 1 and 2 for complex functions such as And-Or/Invert, NORs, etc [1.2]. Under worst-case conditions (125°C, 4.5V and "slow" process conditions) the cross- over point decreases, further favoring BiCMOS. The reason for the improvement offered by the BiCMOS gate is illustrated by a first-order analysis of inverter propagation delay (Td). In a digital inverter Td is inversely proportional to the transconductance (Gm=ΔI/ΔVin) of the gate [1.22]:

$$Td = Cl \; \Delta Vout/\Delta I,$$
$$= Cl \; \Delta Vout/Gm\Delta Vin,$$
$$= Cl/Gm, \text{ when } \Delta Vout = \Delta Vin.$$

A comparison of Gm for an NPN (qIc/kT) and MOS ($(2\mu CoxZ/LId)^{1/2}$) transistor reveals three advantages for the NPN: process and size independence of Gm, and that Gm varies as Ic [1.22]. In order to obtain equal Gm at 10mA the channel width of a 1μm MOSFET would have to be on the order of 100K μm, compared to an emitter area of approximately 60 μm^2. While this analysis is not rigorous --- the minimal extrinsic area required for a MOSFET significantly reduces the area penalty for example (ref. Chapter 6 for a detailed analysis) ---- it does illustrate several important points. The basic BiCMOS inverter is not inherently faster than a CMOS inverter, only more efficient in terms of area and more robust to temperature/voltage and process variations. The area penalty can however be quite severe, as much as a factor of 5 - 7, for worst case process and temperature even in sub-micron technologies. Variations in emitter area are logarithmically reflected in Ic when compared to a ΔVbe, whereas variations in channel length reflect directly when compared to ΔVgs. Finally, because of the $T^{-1.5}$ dependence of mobility, Id is reduced by 40% at 125°C compared to 25°C. These last two attributes make the NPN drive current (and therefore gate delay) more stable than MOSFET drive current.

Sodini, et al have proposed a framework for circuit classification (Table I.1) [1.23]. Bus dominated logic, both memory types, and analog interface circuits can be classified as process technology intensive. As technology improves, performance improves. The remaining categories are design intensive; circuit and

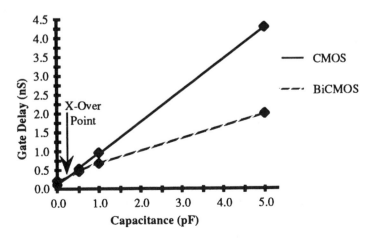

Fig. 1.4 CMOS & BiCMOS inverter delays vs load capacitance for 1μm Lp

device design play a dominant role in determining performance. BiCMOS fits neatly into this framework, analog signal processing being the cross-over point with respect to the circuit categories in which BiCMOS can provide significant performance enhancements. For example, in semi-custom circuits it is difficult to predict nodal capacitance a priori, extensive optimization of critical delay paths is limited by cycle time requirements, layouts cannot be completely optimized for latchup, and I/O flexibility is paramount. Therefore high performance NPNs capable of driving large capacitive loads and providing compatibility with ECL logic levels offer unique advantages. Conversely, in dynamic logic where pre-charging and clocked techniques are used extensively, charge sharing is critical and the need/utility of BiCMOS gates is greatly reduced.

Classification	**Description**	**Example**
Bus Dominated Logic	Loading Dominated by Interconnect	Gate Arrays
Standard Memory	Regular Structures/Technology	SRAM, ROM, PLDs
Special Memory	Specialized Technology	DRAM, EPROM
Analog Interface	"Real World" Interface	Automotive
Analog Signal Processing	Mixed Analog/Digital DSP	Codec/Modems
Dense Static Logic	Combinatorial Logic	Multiplier, ALU
Dynamic Logic	Dynamic Combinatorial Logic	Dynamic ALU

Table I.1 - Circuit Classification & Framework

1.3 BiCMOS Process Technology

BiCMOS technology can be broadly classified into three groups: 1) high performance, 2) low cost, and 3) analog compatible. The high performance (i.e. high speed) BiCMOS technology requires advanced process technology: $\leq 1\mu m$ design rules, thin gate oxides and epitaxial layers, advanced isolation, and 2 - 3 layers of metallization. Circuit density and defect levels must be such that large gate arrays ($\geq 50K$ gates), memories ($\geq 256K$ SRAM) and microprocessors (≥ 32 Bits) can be fabricated with acceptable yields. This high speed category is currently receiving the most attention and the required technology is covered in Chapter 3. The second category, low cost, compromises NPN performance for the sake of minimizing manufacturing costs and maintaining compatibility with existing CMOS processes. Design rules may be scaled more aggressively than for high performance BiCMOS, but the added number of process steps are kept to a minimum. Scaled CMOS is used to increase the circuit density and offset some of the performance loss compared to high performance BiCMOS. Typically an epitaxial layer is not used, and one to two masks added to a baseline CMOS process. In this process, the NPN is usually restricted to the I/O, voltage regulators, and a small number of high capacitance nodes. The third category, analog, also promises tremendous growth potential. Analog requirements differ in that design rules do not have to be scaled as aggressively; 1.5 - 2µm rules suffice for most applications. The challenge of analog is in achieving the proper balance between digital and analog requirements. This stems from the fact that high performance analog technology requirements include higher voltages (10 - 15V), additional components such as precision capacitors and resistors, and high performance PNPs. This implies using thicker gate oxides and epitaxial layers, with a concomitant reduction in digital CMOS and Bipolar performance, and greater process complexity. Analog BiCMOS technology and circuits are covered in Chapters 3 and 8 respectively.

Numerous approaches to implementing a BiCMOS process have been presented [1.13-21]. Simplified sequences and characteristics of the process flows for each of the three categories discussed above (high performance, low cost, and analog compatible) are compared to a typical CMOS and Bipolar logic process flow in fig. 1.5 and Table I.2. All flows are assumed to have double-layer metal, LDD FETs, and twin-tubs.

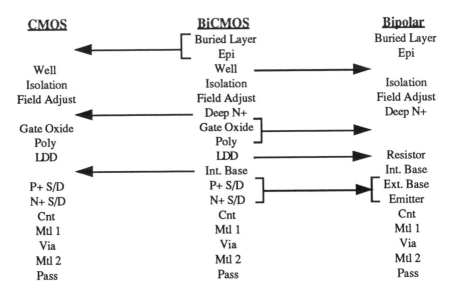

Fig. 1.5 Schematic Representation of BiCMOS Process Steps

From Table I.2 it is seen that mask count is a good indicator of relative process complexity; mask count and total process steps increase by 25% for high performance BiCMOS compared to CMOS. Most of the extra process complexity involved in a BiCMOS is in the front-end --- buried layer and epitaxy. Historically, epitaxy has been an expensive, defect prone step. Recent advances in epitaxy technology and equipment have reduced defect levels to less than $0.25/cm^2$ --- essentially to the background level of incoming substrates [1.2]. The main problem is the interaction of epitaxial growth with the heavily doped N+ buried layer. Careful process optimization is required to avoid increasing defect levels at this step. Still, since NPN transistors typically constitute a low (<15%) percentage of the active chip area in BiCMOS designs, emitter-collector piping defect level requirements are not as stringent as for Bipolar-only technology.

Step	CMOS	BiCMOS			Bipolar
		H. Perf.	**Low Cost**	**Analog**	
Masks	12	15	13	16	13
Etches (RIE)	11	12	11	12	11
Epi	Optional	Required	Optional	Required	Required
Furnace	16	19	16	19	16
Implant	8	12	9	13	7
Metal	2	2	2	2	2
Total	49	61	51	63	50

Table I.2 - Comparison of Process Characteristics [1.2]

1.4 BICMOS Costs

From the comparison in Table I.2 it follows that the BiCMOS chip manufacturers must command a premium for their product in order to pay for the increase in critical process steps and resulting yield loss due to these extra steps. Using high performance BiCMOS as an example, and making several assumptions, a simple cost model can be derived. Assume the yield model $Y = Y_0/(1+ A_{crit} \cdot D_{eff})$, that the increase in critical process steps affect the defectivity level and wafer fabrication costs at a rate 2 - 4 times that of non-critical steps (epitaxy counts as 10X), that $D_{eff}(CMOS) = 1/cm^2$, and that CMOS and BiCMOS are at the same point in the learning curve. Then the increase in cost due to process complexity is ~20% and the BiCMOS defect level is $1.2/cm^2$. This implies that for a die (up to 500 mils on a side) with 90% utilization and critical area, the increase in die costs is 25 to 35%. It follows that high performance BiCMOS must provide sufficient improvement in speed to justify the higher costs compared to CMOS. In a similar fashion BiCMOS must provide performance competitive to Bipolar but at a considerable reduction in power dissipation and cost. A differential in cost over CMOS can be justified if the performance advantage to the end user is demonstrated to be significant -- 25 -50% minimum. On the other hand, in applications where BiCMOS performance does not match Bipolar it must be priced below Bipolar. The rationale supporting this pricing strategy is based on the historical non-linear relationship between chip price and performance.

Epitaxy is often pointed out as being the key step in driving up the cost of BiCMOS technology. Two factors enter into this: yield and equipment cost. As noted earlier it is possible to obtain near defect free epitaxy. The calculations above indicate that yield should not be the main issue. As a matter of fact CMOS vendors are starting to use epitaxy for latchup suppression. The remaining issue is equipment costs. It turns out that the cost for most major pieces of VLSI equipment do not differ significantly (Table 1.3). Taking throughput into account eliminates the cost differential between epitaxial growth and most other steps in the process. Within the limits of this analysis only implantation and metal deposition are cheaper; epitaxy ranks significantly below etch, lithography, planarized inter-layer dielectric (ILD) or contact/via plugs. It follows that adding epitaxy to the process (1 mask/etch, 2 implants/diffusions, and epitaxy) is cheaper than adding a second or third layer of metal (2 critical masks/etches, 1 metal deposition, and 1 dielectric deposition with planarization) to a process. Implantation of a moderately doped buried layer using MeV implantation can be substituted for epitaxy [1.24]. Aside from the process problems and performance loss involved, from Table I.3 it is seen that this approach can be significantly more expensive than growing an epitaxial layer.

Step	Cost ($)*	Throughput (Wfrs/Hr)*	Si Cost ($/Sq In Si)-Hr	Comment
Epitaxy				
AMT 7810	1.00	1.00	3183	1.5µ @ 1070°C
AMT 7010	1.60	2.25	2264	1.5µ @ 1070°C
Implantation				
Nova 80	1.07	5.00	679	5E15 As ~ 4mA
Eaton NV1003	3.33	1.67	6366	1E14As/3.5MeV-.5mA
Etch				
AME 8330	1.07	1.33	2546	1µ Thick AlSi
Lithography				
ASM 2500	1.47	1.25	3735	256K SRAM
Metallization				
Varian 3290	1.07	4.17	815	1µ Thick AlSi
Selective W				
Genus 8710	1.13	0.83	4329	600nm Plug
Planarized ILD				
Precision 5000	1.33	0.67	6366	1µ (Final)

Table 1.3 - Relative Cost of VLSI Equipment [1.2]
* Normalized to the AMT 7810

1.5 Applications

BiCMOS fills the market niche between very high speed, but power hungry Bipolar ECL and the very high density, medium speed CMOS. In a sense, BiCMOS technology can be viewed as the AND function of high speed and high density. This implies that the BiCMOS marketing challenge is to position the technology so that market share is taken away from both Bipolar ECL and CMOS, while opening up new markets. The Bipolar ECL market has historically pursued speed "at all cost". Before this market segment is addressed BiCMOS circuit implementations will have to either match the speed of the best available Bipolar circuits at a lower power or, vice versa, exceed the speed of the Bipolar circuit at the same power. It should be clear that when the power budget is unconstrained, a Bipolar technology optimized for speed will almost always be faster than a BiCMOS technology. However, in applications where a finite power budget exists the ability to focus power where it is required usually allows BiCMOS speed performance to surpass that of Bipolar [1.25]

In the past BiCMOS gate delays have been 2 to 3 times that of high performance ECL: 450ps vs 200ps for 1.5μ BiCMOS and Bipolar 2-Input gates (NANDs/NORs) respectively [1.17]. ECL series gating can accentuate this difference for more complex functions. In order to overcome these drawbacks BiCMOS NPN performance has to be improved significantly and mixed circuits (ECL and CMOS) have to be used effectively. These deficiencies have been recently addressed [1.26-1.27]. In $1\mu m$ technology BiCMOS gate delays of <250psec for a 2-Input NAND and sub-100ps ECL gate delays have been demonstrated [1.27]. This rivals gate delays in Bipolar ECL. It follows that by utilizing ECL circuitry only where necessary, equivalent, or even superior, speed performance can be provided at a fraction of the power.

BiCMOS technology is especially well suited for I/O intensive applications. ECL, TTL, and CMOS input and output levels can easily be generated with no speed or tracking consequences. CMOS to true ECL (-5.2V Vee) translation takes ~500ps, whereas ECL to CMOS translation takes 2 - 3ns in 2μ BiCMOS. Translation from pseudo ECL (+5V Vcc) to CMOS levels takes significantly less time, ~ 500ps. TTL output cells that can handle up to 24mA Iol can also be switched in 2 - 3 ns. High drive CMOS outputs suffer from extremely sharp rise and fall edges. This results in large overshoots/ground bounce in the system. A great deal of design effort has gone into reducing these edges, but the penalty in speed and/or area is severe. Conversely, TTL Bipolar output edges are not a direct function of the output devices, but rather of the phase splitter skew, collector resistance, Schottky clamp, and squaring network. This results in slower edge rates while maintaining fast propagation delays. This is becoming of critical importance from a systems perspective as internal CMOS gate delay is improved. The rail to rail swings and unterminated environment used in CMOS makes output switching above 33MHz difficult. BiCMOS's inherent compatibility with ECL or TTL levels provides an ideal solution to these problems.

Arguments are made that when die size gets large enough and CMOS gets small enough all circuit problems will be solved. The BiCMOS advantages of significantly higher system level performance than CMOS at a lower total cost have to be convincingly demonstrated. Several factors indicate that this is taking place. First, the capital costs of investing in continually smaller ($<1\mu$) CMOS technology increase exponentially [1.28], while the requirement of lower power supplies for 0.5μ CMOS degrades performance [1.29]. Since BiCMOS does not have to be scaled as aggressively as CMOS, capital costs are not as high (existing fabs can be utilized) and the power supply can be maintained at 5V. This translates into lower chip cost with better performance and 5V compatibility. Larger chips impose severe performance penalties due to simultaneous switching noise, internal clock skews, and high nodal capacitances

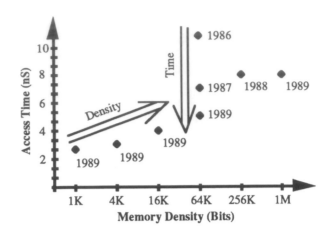

Fig. 1.6 BiCMOS ECL Compatible Memory Density vs Access Time

in critical paths; all factors for which BiCMOS has been demonstrated to be superior to CMOS. Finally, BiCMOS technology can take advantage of any advances in CMOS and/or Bipolar technology, greatly accelerating the learning curve normally associated with new technologies. A few examples best illustrate these points.

1.5.1 Memories

In memories BiCMOS provides faster access time at the same power as all-Bipolar or the same access time at lower power. In addition, the CMOS memory cells used in BiCMOS memories have superior stability, better alpha resistance, and smaller area than an equivalent bipolar cell. Compared to CMOS, BiCMOS provides higher speed at equivalent densities plus ECL I/O capability. Since they share the same memory cell, die area for BiCMOS and CMOS memories are approximately the same. With recent introductions, BiCMOS memories now covers the full density spectrum from 1K to 1M at speeds below 10ns (fig. 1.6). The small, 1K to 4K, BiCMOS SRAMs utilize high speed bipolar ECL for decoding, sensing, and output drivers, and a BiCMOS memory cell. A symmetrical read/write access time of 3ns is achieved while consuming only 180mA [1.30]. In high density memories, 8ns 256K and 1M bit BiCMOS ECL SRAMs have been demonstrated [1.31-1.34]. These SRAMs provide bipolar speeds at densities which previously could only be achieved in CMOS.

Deciding which gates to implement in Bipolar vs CMOS vs BiCMOS is an exercise is power/speed budget management. An example of how a typical TTL BiCMOS SRAM could be implemented is shown in fig. 1.7. Since NPNs provide large output drive, BiCMOS logic gates are effectively applied at high capacitive nodes. This occurs in decoders, wordline drivers, write drivers and output buffers. Control logic functions with small fan-out use CMOS. Sense amplifiers, which require high input sensitivity, typically uses pure bipolar.

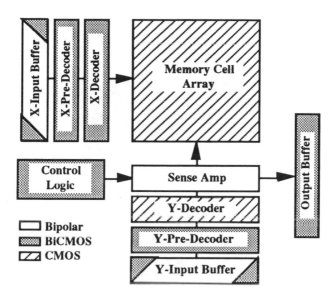

Fig. 1.7 TTL SRAM Circuit Blocks as Function of Technology

Bipolar differential pairs provide the high gain and input sensitivity required to quickly sense small differences in bit line swings. A potential problem is that as memory cell size is reduced it becomes difficult to fit NPNs in the cell pitch. In ECL BiCMOS SRAMs speed is even more critical and a larger percentage of circuits are implemented in pure bipolar. The translation from ECL to CMOS logic levels is postponed until after decoding, and bipolar ECL gates are used for input buffers, decoding, sensing, and output buffers [1.35].

1.5.2 Semi-Custom [1.2]

In semi-custom applications BiCMOS technology offers performance and I/O flexibility advantages over CMOS. With respect to Bipolar semi-custom offerings, BiCMOS provides lower power and higher density at nearly the same speed level. A schematic illustration of gate delay vs power is provided in fig. 1.8. While most of the semi-custom applications to date have been in gate arrays, standard cell BiCMOS benefits from the same circuit and architectural possibilities as do arrays. However, due to the greater degrees of freedom in cell design, many novel circuit structures can be created. This trend will be seen at the megacell level, where gates or even functions can be exclusively CMOS, ECL or BiCMOS. These cells will then be merged at the final level to perform the necessary function.

A wide range of gate array architectures and circuit techniques for the development of high performance and high density gate arrays have been proposed. Architectures which use NPNs in internal cells and I/O versus just in the I/O have both been demonstrated [1.36]. The tradeoff is between performance/ design ease and density. Initially, the internal core section was CMOS-only with

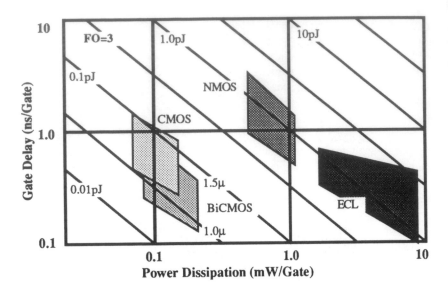

Fig. 1.8 Gate Array Cell Speed vs Power

the I/O employing the Bipolar devices as required. This made for an area efficient core, but did not take full speed advantage of the bipolars. Advocates of the "I/O only" approach propose BiCMOS as a solution to the LΔI/Δt problem. This approach only makes sense for a low cost BiCMOS implementation, otherwise very poor utilization is being made of the technology investment. This is especially true if CMOS performance is compromised by the addition of the NPNs. The next evolution incorporated the Bipolars into the basic cell within the array [1.36]. This provides excellent drive capability to the CMOS functions while maintaining low power consumption, but at the expense of 20% increase in cell area. Since the Bipolar is only required ~60% of the time, it follows that this architecture tends to waste chip area [1.2]. The current trend focuses on the use of Bipolar devices in large macro function (adder, register, ST flip-flops, etc.) [1.2]. This makes more efficient use of the Bipolar devices in an average option and is the approach that will be taken for very high (30 - 100K gates) density arrays. Depending on the architecture, the area overhead for the macrocell approach can be 5% or less. A new class of arrays, application specific arrays, with embedded architectures (RAM, ECL core, analog) have also recently emerged (Chapter 5).

1.5.3 Microprocessors

BiCMOS can significantly enhance microprocessor (μP) performance by reducing delays through critical paths, reducing clock skews, and increasing I/O throughput. No major supplier has introduced a BiCMOS processor, but recent announcements from Texas Instruments, Motorola, Fujitsu, and Hitachi imply that several are on the way [1.37]. Initial applications of BiCMOS may be in RISC μPs as opposed to the more traditional CISC processors. This may be due to the higher clock speeds that RISC μPs must operate at in order to remain competitive with CISC processors. The tradeoff between CISC and RISC is being reduced to chip size and availability/speed of on-chip memory [1.38]. The ability to integrate upwards of 512K of on-chip 6-8ns SRAM and then communicate at greater than 70 MHz with an external bus will ultimately determine the performance limit of circa 1990 - 1991 μPs. These two areas, memory and I/O, are applications at which BiCMOS excels.

Because of the higher clock rates required, RISC processors are leading candidates for BiCMOS integration [1.38]. The simpler instruction set used in RISC processors imply that fast access to on- and off- chip cache is necessary. It also implies that to eliminate "wait states" faster memory is required. To realize an 80 - 100 mips machine requires 6 - 8ns SRAM [1.38]. This speed cannot be sustained in CMOS over worst-case temperature and voltage. A number of companies are trying to adapt Bipolar technology to RISC μPs, but the inability to integrate the required amount of on-chip memory and Bipolar's high power dissipation make this approach impractical. BiCMOS can: 1) utilize CMOS design methodologies used to design previous generations of μPs while also providing ECL data paths as necessary, 2) integrate large fast on-chip RAM, and 3) keep power manageable by focusing it where required to optimize the speed-power tradeoff. Given these advantages it is not surprising that BiCMOS versions of SUN's Sparc and MIP's R3000 RISC μPs are in design [1.37].

One example of a BiCMOS implementation of a μP has been carried out by Hitachi. Their approach was to utilize a 1.3μm BiCMOS standard cell library to realize a high speed processor. Various building blocks, including 64 X 2K of μcode ROM, 32-bit ALU, cache memory, 64-bit multiplier, PLA, and register files were integrated in a test chip. The result was a 25ns (worst case) machine cycle time [1.18]. In a scaled and expanded version of this early work, a 70MHz 32 bit microprocessor was designed in 1.0μm BiCMOS [1.39]. That chip contained over half a million transistors, only 1.5% of which were NPNs. The bipolar transistors were used in critical speed paths and in sense circuits. This is reminiscent of the early implementation of CMOS μPs, in which only a few true CMOS gates were used.

1.5.4 Analog/Digital Systems [1.19]

There are a large number of applications for BiCMOS technology which will result in a single chip straddling the analog-digital boundary. In such circuits the greater part of the silicon area, utilizing CMOS, will be used for the digital signal processing of signals. A much smaller portion will be devoted to the essential analog processing needed in order to interface to the outside world. The analog-digital boundary no longer a matter of chip partitioning, the circuit designer can choose the most suitable domain for each part of the required solution. This benefit leads to the most efficient use of silicon in a complex system.

Bipolar technology had long been the domain of classical analog circuits. The advent of switched capacitor circuit technology in the late 1970s caused a shift to analog CMOS for telecommunications, data conversion, and filter applications. Because of the advantage of precision monolithic filtering and charge redistribution techniques, MOS analog support circuits were developed despite the fact that Gm was an order of magnitude lower, matched pair offset an order of magnitude higher, and 1/F noise higher with an order of magnitude higher corner frequency than equivalent Bipolar technology. Scaling the FETs does not help significantly since even though Ft improves with lower Leff, Rout and consequently the open loop circuit gain suffers. Therefore Bipolar technology holds numerous advantages in areas of classical analog circuits such as operational amplifiers and regulators (See Chapter 8). BiCMOS bridges the gap allowing process and circuit boundaries to be crossed.

A list of the circuit design advantages for BiCMOS compared to a Bipolar or CMOS processes are: 1) minimal current drain, automatically scalable for low and medium speed digital control logic, 2) high impedance inputs (FETs) for sample and hold applications, 3) >1GHz toggle frequency, 4) low 1/F noise, 5) high gain Bipolars, 6) low input offsets voltage for differential pairs, 7) "zero offset" analog switches, 8) gain-bandwidth product extended, and 9) good voltage references. These design advantages can be used in communication circuits, such as a single chip direct divide synthesizer. A further example of BiCMOS's flexibility is in the fast growing realm of oversample noise shaping and predictive coders [1.19]. The circuit requirements for the coder itself demands low noise, high speed linear devices, but these must be followed by a complex digital filter. Thus advanced coders of this type would reside in two packages unless BiCMOS is employed. The central office subscriber line card is another environment where the advantages of BiCMOS can be put to good use. The traditional Bipolar SLIC functions can be combined with the DSP COFIDEC functions in a single package giving greater flexibility and controllability to the system [1.19]

1.6 Projections

For a given technology level, BiCMOS is now accepted as providing enhanced performance over CMOS by a factor of 1.5 - 2.0X. A 0.8μ BiCMOS technology exceeds the performance of a sub-0.5μm CMOS technology in conventional applications such as SRAMs, gate arrays, and microprocessors. The area that still solicits controversy is the role of BiCMOS in a reduced power supply environment. A 3.3V power supply is incompatible with the standard ECL interface and at 3.3V the performance of the conventional BiCMOS buffer is severely degraded [1.27, 1.40]. Two approaches are being taken to resolve these problems. The first is to provide on-chip voltage regulation in order to operate sub-0.5μm CMOS at lower electric fields [1.41]. This allows the I/O interface to remain at standard levels while providing for greater density and lower power dissipation internally. The second approach is to replace the conventional BiCMOS buffer and ECL gate with new logic gates capable of operating at 3.3V or less [1.42-1.43]. It follows that BiCMOS technology will extend the use of the standard TTL and ECL interfaces into the sub-micron regime. This will simultaneously allow for greater performance, higher densities, and lower power; all at the industry standard 5V. Whether BiCMOS will become the technology driver of the 1990s as proposed by the SRC BiCMOS roadmap committee remains to be seen [1.44].

One area in which a great deal of investigation is still required is CAD tools for circuit design. Even though mixed circuit simulation, design rule checking and layout verification is not a problem, there are currently no available logic simulators or compilers that can handle mixed CMOS and ECL logic circuits. Even mixing CMOS and BiCMOS circuits at the logic level can be a problem. As BiCMOS increases in importance, especially for semi-custom design, software tools currently available for CMOS will be adapted to BiCMOS. This is an area in which CMOS, because of marketshare and design ease, has a clear advantage over both bipolar and BiCMOS technology. Still, as the capability of new cell generators improve, CAD will become less of an issue.

1.7 Summary

BiCMOS technology significantly enhances speed performance while incurring a negligible power or area penalty. Thus BiCMOS can provide applications with CMOS power and densities at speeds which were previously the exclusive domain of bipolar. This has been demonstrated in applications ranging from static RAMs to gate arrays to microprocessors. Thus the concept of a "system on a chip" becomes a reality with BiCMOS. The main disadvantage of BiCMOS is greater process complexity, which results in a 1.25-1.4X increase in die costs over conventional CMOS. Taking into account packaging costs, the total manufacturing costs of supplying a BiCMOS chip ranges from 1.1 - 1.3X that of CMOS. The extra costs incurred in developing a BiCMOS technology is more then offset by the fact that the enhanced chip performance obtained extends the usefulness of manufacturing equipment and clean rooms by at least one technology generation. BiCMOS is now being demonstrated at 0.5μm. This will extend the conventional |5V| TTL and ECL interfaces another generation of technology, thereby maintaining the investment in 5V systems. While BiCMOS will not necessarily displace CMOS as the technology of the 1990s, it will have a significant impact on the Integrated Circuit industry.

References

1.1] A.R. Alvarez, P. Meller, B. Tien, "2 μm Merged BIMOS Technology," 1984 Int. Electron Devices Meeting, pp. 420-424.

1.2] A.R. Alvarez, D.W. Schucker, "BiCMOS Technology for Semi-Custom Integrated Circuits," Cust. Int. Cir. Conf., pp. 22.1.1 - 22.1.5, 1988.

1.3] H. G. Lin, J.C. Ho, R.R. Iyer, K. Kwong, "Complementary MOS-Bipolar Transistor Structure," IEEE Trans. Electron Devices, Vol. ED-16, No. 11, Nov.

1969, pp. 945-951.

1.4] M.A. Polinsky, O.H. Schade, J.P. Keller, "CMOS-Bipolar Monolithic Integrated Circuit Technology," 1973 Int. Electron Devices Meeting, pp. 229-231.

1.5] J. D. Plummer, J.D. Meindl, "A Monolithic 200-V CMOS Analog Switch,"J. Solid State Circuits, SC-11, No. 6, Dec. 1976, pp. 809-817.

1.6] S. Davis, "Simplified Driver, New Applications Spur Plasma-Panel Usage," Electronic Design News, Sept. 20, 1979, pp. 51-55.

1.7] B. Holland, D.P. Peppenger, "Single-Chip Linear Regulator Handles 125V I/O Differential," Electronic Design, Sept. 17, 1981, pp. 129-133.

1.8] T.E. Ruggles, G.V. Fay, "Mixed Process Puts High Power Under Control," Electronic Design, Vol. 30, No. 7, March 31,1982, pp.69-77.

1.9] A.R. Alvarez, R.M. Roop, K.I. Ray, G. R. Gettemeyer, "Lateral DMOS Transistor Optimized for High Voltage BiCMOS Applications," 1983 Int. Electron Devices Meeting, pp. 420-423.

1.10] S. Lytle, R. Roop, D. Cave, D. Hughes, W. Gegg, A.R. Alvarez, "Power BiCMOS - A Versatile IC Technology for Switching and Regulation Applications", 1984 Custom Integrated Circuits Conference, pp. 51-56.

1.11] E.J. Wildi, T.P. Chow, M.S. Adler, M.E. Cornell, G.C. Pifer", New High Voltage IC Technology", 1984 Int. Electron Devices Meeting pp. 262-265.

1.12] R.T. Gallager, "Single Chip Carries Three Technologies," Electronics Week, Dec. 10, 1984, p. 28.

1.13] H. Higuchi, G. Kitsukawa, T. Ikeda, Y. Nishio, "Performance and Structures of Scaled-Down Bipolar Devices Merged with CMOSFETs," Int. Electron Devices Meeting, p. 694- 697, 1984.

1.14] J. Miyamoto, S. Saitoh, H. Momose, H. Shibata, K. Kanzaki, S. Kohyama," A 1.0μm N-Well CMOS/ Bipolar Technology For VLSI Circuits", 1984 Int. Electron Devices Meeting, pp. 63 - 66.

1.15] B. Bastani, C. Lage, L. Wong, J. Small, R. Lahri, L. Bouknight, T. Bowman, J. Manoliu, P. Tuntasood, "Advanced 1μ BiCMOS Technology for High Speed 256K SRAMs," 1987 Symp. on VLSI Technology, pp. 41 - 42.

1.16] N. Tamba S. Miyaoka, M. Odaka, M. Hirao, K. Ogiue, K. Tamada, T. Ikeda, H. Higuchi, H. Uchida, "An 8ns 256K BiCMOS SRAM," 1988 Int. Solid State Circuits Conference, pp. 184 - 185.

1.17] A.R. Alvarez, J. Teplik, D.W. Schucker, T. Hulseweh, H.B. Liang, M. Dydyk, I. Rahim, "Second Generation BiCMOS Gate Array Technology," 1987 Bipolar Circuits and Technology Meeting, pp. 113 - 117.

1.18] T. Hotta, I. Masuda, H. Maejima, A. Hotta, "CMOS/Bipolar Circuits for 60MHz Digital Processing," 1986 Int. Solid State Circuits Conference, pp. 190-191.

1.19] A.R. Alvarez, J. Teplik, H.B. Liang, T. Hulseweh, D.W. Schucker, K.L. McLaughlin, K.A. Hansen, B. Smith, "VLSI BiCMOS Technology and Applications," 1987 International Symp on VLSI Technology, Systems, and Applications, pp. 314 - 319.

1.20] M.P. Brassington, M. El-Diwany, P. Tuntasood, R.R. Razouk, "An Advanced Submicron BiCMOS Technology for VLSI Applications," 1988 Symp. on VLSI Technology, pp. 89 - 90.

1.21] M. Kubo, "Perspective on BiCMOS VLSIs," Symp. VLSI Technology Dig. Tech. Papers, pp. 89 - 90, 1987. Also in IEEE J. Solid State Circuits, Vol. SC-23, No. 1, Feb. 1988, pp. 5-11.

1.22] B.L. Morris, "BiCMOS Digital Design Techniques and Applications," IEDM

BiCMOS Technology and Design Short Course, 1987.

1.23] C. Sodini, S.S. Wong, P.K. Ko, "A Framework to Evaluate Technology and Device Enhancements for MOS Integrated Circuits," IEEE J. Solid State Circuits, Vol. SC-24, No. 1, Feb. 1989, pp. 118-127.

1.24] H.J. Bohm, L. Bernewitz, W.R. Bohm, R. Kopl, "Megaelectronvolt Phosphorus Implantation for Bipolar Devices," IEEE Trans. Electron Devices, Vol. ED-35, No. 10, Oct. 1988, pp. 945-951.

1.25] R.C. Lutz, Aspen Semiconductor Corp., Private Communication, 1988.

1.26] T.Y. Chiu, et al, "A High Speed Super Self-Aligned Bipolar-CMOS Technology," Int. Electron Device Meeting, pp. 24 - 27, 1987.

1.27] T. Yuzuriha, T. Yamaguchi, J. Lee, "Submicron Bipolar-CMOS Technology Using 16 GHz Ft Double Poly-Si Bipolar Devices," Int. Electron Device Meeting, pp. 748-751, 1988

1.28] K. Shibayama, Y. Akasaka, "Laboratory and Factory Automation for ULSI Development and Mass Production," Int. Electron Device Meeting, pp. 736-739, 1988.

1.29] R.A. Chapman, C.C. Wei, D.A. Bell, S. Aur, G.A. Brown, R.A. Haken, "0.5 Micron CMOS for High Performance at 3.3V," Int. Electron Device Meeting, pp. 52-55, 1988.

1.30] B. Cole, "Aspen Shows BiCMOS Can Yield Fast SRAMs," Electronics pp. 88, Feb. 1989.

1.31] H.V. Tran, D.B. Scott, K. Fung, R. Havemann, R.E. Eklund, T.E Ham, R.A. Haken, A.H. Shah, "An 8ns Battery Backup Submicron BiCMOS 256K ECL SRAM," Int. Solid State Circuits Conf. Dig. Tech. Papers, pp. 188 - 189, 1988.

1.32] R.A. Kurtis, D.D. Smith, T.L. Bowman, "A 12ns 256K BiCMOS sRAM," Int. Solid State Circuits Conf. Dig. Tech. Papers, pp. 186 -187, 1988.

1.33] M. Matsui, H. Momose, Y. Urakawa, T. Maeda, A. Suzuki, N. Urakawa, K. Sato, K. Makita, J. Matsunaga, K. Ochii, "An 8ns 1Mb ECL BiCMOS SRAM," Int. Solid State Circuits Conf. Dig. Tech. Papers, pp. 38 39, 1989.

1.34] H.V. Tran, K. Fung, D. Bell, R. Chapman, M. Harward, T. Suzuki, R. Havemann, R.Eklund, R. Fleck, D. Le, C. Wei, N. Iyengar, M. Rodder, R.A. Haken, D.B. Scott, "An 8ns 1MB ECL SRAM with a Configurable Memroy Array Size," Int. Solid State Circuits Conf. Dig. Tech. Papers, pp. 36 - 37, 1989.

1.35] M. Suzuki, S. Tachibana, A. Watanabe, S. Shukuri, H. Higuchi, T. Nagano, K. Shimohigashi, "A 3.5ns/500mW 16Kb BiCMOS ECL RAM," Int. Solid State Circuits Conf. Dig. Tech. Papers, pp. 32-33, 1989.

1.36] P.T. Hickman, F. Ormerod, D.W. Schucker, "A High Performance 6000 Gate BIMOS Logic Array," Custom Int. Circuit Conference, pp. 562 - 564,1986 .

1.37] B. Cole, "Now There's More Than Just Raw Speed in The ECL Arena," Electronics, pp. 84 - 87, Feb. 1989.

1.38] P. Bosshart, "Introduction to Microprocessor Design," Int. Electron Device Meeting Short Course, 1988.

1.39] T. Hotta, T. Bandoh, A. Hotta, T. Nakano, S. Iwamoto, S. Adachi, "A 70Mhz 32b Microprocessor with 1.0μm BiCMOS Macrocell Library,," Int. Solid State Circuits Conf. Dig. Tech. Papers, pp. 124-125, 1989.

1.40] H. Momose et al, "0.5μm BiCMOS Technology," Int. Electron Device Meeting, pp. 838 - 840, 1988.

1.41] H. Fukuda, S. Horiguchi, M. Urano, K. Fukami, K. Matsuda, N. Ohwada, H. Akiya, "A BiCMOS Channelless Masterslice with On-Chip Voltage

Converter," Int. Solid State Circuits Conf. Dig. Tech. Papers, pp. 176-177, 1989.

1.42]   W. Heimsch, B. Hoffmann, R. Krebs, E. Muellener, B. Pfaeffel, K. Ziemann, "Merged CMOS/Bipolar Current Switch Logic,"Int. Solid State Circuits Conf. Dig. Tech. Papers, pp. 112-1133, 1989.

1.43]   Y. Nishio, F. Murabayashi, S. Kotoku, A. Watanabe, S. Shukuri, K. Shimohigashi, "A BiCMOS Logic Gate with Positive Feedback," Int. Solid State Circuits Conf. Dig. Tech. Papers, pp. 116-117, 1989.

1.44]   J.A. Coriale, W.C. Holton, (Eds), "Submicron BiCMOS Technology for the 90s," SRC Topical Research Conference, Dec. 10 - 11, 1087. N. Anatha "BiCMOS Roadmap".

Chapter 2

Device Design

J. Teplik (Motorola Inc.)

2.0 Introduction

The increasing acceptance of BiCMOS as a viable technology has brought to the forefront new challenges for the device physicist/designer. Ultimately, the device designer wishes to optimize the performance of the transistors while simultaneously maintaining process simplicity. The challenge to achieve this in the BiCMOS environment is heightened because the process requirements for the MOSFET and bipolar transistors often conflict with one another. Most of the compromises involved are determined in the design of the front-end of the process. However, since the MOSFET and BJT characteristics are strongly coupled, optimization of both devices can only occur at the expense of increased process complexity and the associated manufacturing cost. A thorough device design approach, coupled with the application of a statistically-based device design methodology, becomes critical for evaluating both performance tradeoffs and manufacturability implications.

Device design is one piece of an overall technology development (fig. 2.1). Circuit/system specifications provide the baseline for the initial device targets. Process module considerations and compatibility with existing processes impose further constraints on the development of devices. Initial integration is a merging of these requirements; feedback from physical, electrical, and circuit evaluations permit refinement of the device performance through the use of process and device simulation tools. The device design methodology provides a vehicle for the quantification of the region of optimum performance in terms of the process variables, as well as the evaluation of device characteristic sensitivities to manufacturing tolerances.

In this chapter, the key MOSFET and BJT parameters are discussed to lay a foundation for examining the tradeoffs inherent to BiCMOS device development. Typical values are cited throughout this discussion, to provide the reader with a quantitative baseline. Using a device design flowchart developed by applying response surface methods, the front-end options are discussed in some detail, since most of the compromises and necessary optimization lies here. A device

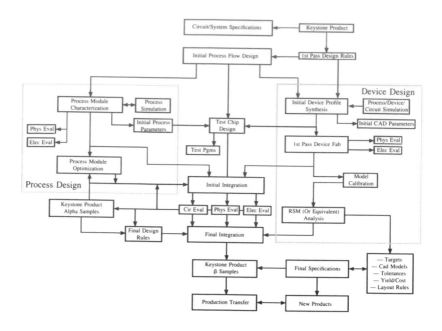

Figure 2.1. Technology development flowchart, integrating process, device, and design considerations. From [2.1]. © 1987 IEEE.

design methodology is introduced after the distinction between synthesis and analysis has been clarified. The motivation for, and application of response surface methods coupled with statistical techniques is described, followed by a discussion of the analysis and implications regarding target and manufacturing optimization. A methodology of coupling device model parameters to circuit design considerations is presented. Finally, scaling issues for both MOS and NPN devices are reviewed and their implications regarding future technology development are discussed.

2.1 Device Issues

2.1.1 MOS Devices

In examining the intrinsic transistor (i.e. excluding contacts, metal, or poly interconnects), MOS device development can be divided into channel design and source/drain design issues (fig. 2.2). Ion implantation is used to control the dopant profiles for these two regions; this allows independent tailoring of each. However, high-temperature processing steps in the back-end of the BiCMOS process may be common to both profiles. Therefore the parameters of the channel and source/drains are somewhat coupled to one another.

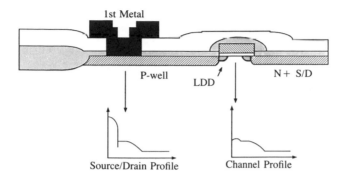

Figure 2.2. MOSFET design divided into channel and source/drain issues. From [2.1]. © 1987 IEEE.

Channel Considerations. To the first order, the channel implant determines the threshold voltage, and so is chosen to meet this basic requirement. The threshold voltage, which is typically 600 to 900mV, is set by noise margin, speed, and leakage considerations. The background concentration, which is set by either the well doping or epi concentration, may also impact V_T, but this is usually a second-order effect (fig. 2.3). The background doping *will* determine the body effect, which is a measure of the sensitivity of the device threshold to changes in substrate bias. The combination of the channel implant and to a lesser degree, the background doping, will effect the transconductance, or gain of the device, through the mobility. Although the mobility in the inversion layer is not determined by the surface doping density for doping less than 1E17 cm^{-3}, it is proportional to the effective electric field in the inversion layer, which increases with doping [2.2].

Since the transconductance of a MOSFET is inversely proportional to the

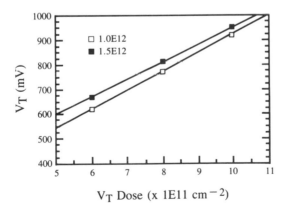

Figure 2.3. NMOS threshold as a function of channel implant dose, for 2 different P-well doses.

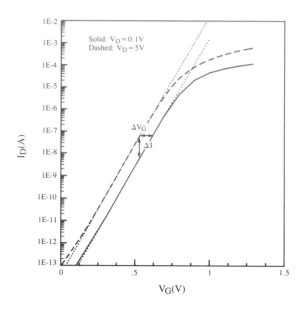

Figure 2.4. Simulated subthreshold characteristics of a short-channel PMOS device as a function of drain voltage. Leff=0.7μm.

channel length, it is advantageous to decrease the gate length to improve the current drive. This leads to the well-known short-channel effects, in which the gate no longer has full control of the channel. The close proximity of the drain to the channel region causes a significant amount of the bulk charge in the channel to be terminated by the drain depletion charge; more of the charge on the gate is available to terminate on charge in the inversion layer, and thus the current is greater than predicted by first-order theory. This description is known as the charge-sharing concept [2.3]. An alternate explanation is that the close proximity of the drain to the source allows the drain to reduce the potential barrier at the source end of the channel, which increases the flow of carriers into the channel. This is known as drain-induced barrier lowering (DIBL) [2.4]. As the drain voltage increases, the effect becomes more significant. Ultimately, an increase in drain voltage results in punchthrough.

DIBL is most readily evaluated by examining the subthreshold characteristics as a function of drain voltage, as shown in fig. 2.4. The significance of short-channel effects can be quantified by measuring the shift in the gate voltage ΔV_G for a fixed drain current, usually 1nA x W/L (fig. 2.4). This gives a good first-order estimate of the expected shift in threshold voltage as a function of drain bias. It is convenient to report the gate shift in a normalized form $\Delta V_G/\Delta V_D$. An acceptable value is approximately 25mV/V, depending on the

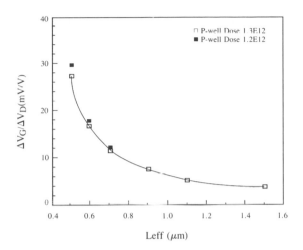

Figure 2.5. Simulated short channel effects as a function of channel length.

circuit application and specific leakage requirements. $\Delta V_G/\Delta V_D$ is a strong function of channel length (fig. 2.5) and must be guaranteed to at least 3 sigma below the target channel length.

An alternate means of characterizing the short-channel effects is to measure the change in drain current as a function of drain voltage for a fixed gate voltage, which is defined as ΔI (fig. 2.4). Because of its simplicity, ΔI (in decades) is easily implemented on automated test equipment.

Short-channel effects can be minimized by increasing the background concentration, but this is at the expense of increased body effect and junction capacitance (fig. 2.6). To minimize the impact on speed, the shift in the threshold due to the body effect is kept below 900mV. An alternative method to controlling short-channel effects is to place a second implant in the channel region below the first implant. Through proper design of this deeper implant, the impact on body effect and capacitance is minimized. This requires an additional mask and so increases process complexity. Another possibility is to decrease the gate oxide thickness; the thinning of the gate places the gate electrode closer to the channel and therefore gives it greater control of the channel. However, thinning the gate oxide increases the maximum electric field according to $E_p = (V_D - V_{Dsat})/\sqrt(3t_{ox}x_j)$ [2.5]. This results in increased hot carrier generation, which is typically monitored as substrate current: $I_{sub}=CI_Dexp(\beta/E_p)$, where C and β are constants. These effects can be compensated for by careful design of the source/drain regions.

In addition to determining the threshold voltage and the short-channel characteristics, the channel profile design also determines the leakage of the

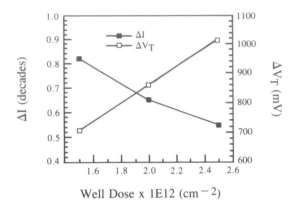

Well Dose x 1E12 (cm^{-2})

Figure 2.6. Short-channel effects (ΔI) are controlled by increasing the well dose at the expense of increased body effect. ΔV_T is the shift in V_T as V_{BG} is changed from 0 to 5V.

device, which is a special concern in random access memories (RAMs). In the subthreshold region, the swing S is used to quantify the amount that the gate voltage must be reduced to decrease the current by one decade [2.6]. Acceptable values for S are 95mV/decade or less. Since the swing is proportional to the depletion capacitance and inversely proportional to the oxide capacitance, the swing improves (i.e. gets smaller) with a decrease in the dose or energy of the channel implant (fig. 2.7). Decreasing the oxide thickness is also beneficial to the subthreshold characteristics, as shown in fig. 2.7.

The preceding discussion has assumed that the MOSFETs are surface-channel devices. This is the case when the polysilicon gates are N-doped and P-doped for the NMOS and PMOS devices respectively. However, this scheme increases the process complexity of the interconnects. By using a "single-flavored" N-doped poly, the process is simplified, and the penalty of higher resistance associated with P-doping (due to lower mobility) is eliminated. However, the work function of the N-type poly requires that the PMOS channel be counterdoped to achieve the desired threshold (approximately -700mV). This results in a buried-channel device. For such a device, the transverse electric field penetrates deeper into the silicon, and the minimum potential barrier is subsurface (fig. 2.8). Although this typically results in a 15% improvement in mobility and a higher I_{Dsat} compared to a surface channel device, DIBL is much more severe; ΔV_G in the subthreshold region can be twice as high or more compared to a surface channel device. Therefore, special processing such as punchthrough implants or retrograde wells must be used to control the short-channel behavior. An alternative is to use refractory metals such as molybdenum for the gate material. This gives work function values for both the NMOS and PMOS that are close to the bandgap and thus results in symmetrical devices. However, the inability to prevent punchthrough for devices shorter than 0.5µm makes using

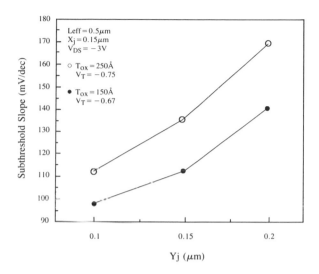

Figure 2.7. Simulated PMOS subthreshold slope versus counter-doping junction depth into the silicon Y_j for different oxide thicknesses. From [2.7]. © 1984 IEEE.

refractory metals questionable [2.9].

The impact of the channel implant dose and energy, background concentration, gate oxide thickness, and back-end Dt on the key MOSFET parameters is summarized in Table 2.1. Because the lightly-doped side of a junction determines the capacitance and the breakdown voltage to the first order, these parameters will also be controlled by the background and channel profiles.

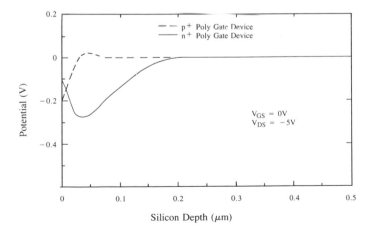

Figure 2.8. Simulated electrical potential perpendicular to the surface in the channel region. N+ poly is buried-channel device. From [2.8]. © 1985 IEEE.

| Output | ↓ tox | V$_T$ Adjust | | Post Implant | | ↓ xj |
		↑ Dose	↓ Energy	↓ Dt	↑ Nbc	
V$_T$	↓	↑	↑	↑	↑	↑
Body Effect	↓	↑	↓	↓	↑	↑
G$_m$	↑	↓	↓	↓	↓	↓
Subthreshold Swing	↓	↑	↓	↓	↑	↑
ΔI	↓	↓	↑	↑	↓	↓
C$_{GD}$	—	↑	—	↓	↓	↓
R$_S$	—	—	—	↑	—	↑
C$_j$	—	↑	↓	*	↑	↓
HCI	↑	↑	↑	↑	↑	↑
Breakdown Voltage	↓	—	—	↓	↓	↓

* depends on Nbc profile

Table 2.1. Primary process parameters, and the resulting impact (increase or decrease) on the key electrical characteristics of a MOSFET.

Therefore, the parasitic effects associated with the source/drains will be coupled to the channel design.

The aforementioned effects will set the constraints on the channel and background profiles. The peak background doping will be limited by the body effect and capacitance requirements, while the lower boundary on doping will be set by the short-channel and punchthrough considerations. The peak channel doping will be constrained by subthreshold swing limits and mobility considerations, while the lower limit will be restricted by surface short-channel effects. The range of acceptable channel dose will be further constrained by the threshold voltage requirement. For high-performance BiCMOS, typical background doping is between 1E16 and 3E16 cm^{-3}, while channel doping can be as high as 1E17 cm^{-3}.

Source/drain Design Considerations. The second half of Table 2.1 summarizes the important issues for source/drain design. By minimizing the junction depth of the source/drains, better short-channel effects and minimal gate overlap capacitance is achieved; this is at the expense of worse hot-carrier injection (HCI) and higher source/drain sheet rho, which may limit the use of diffused regions as interconnects.

Hot carrier effects, which primarily result in degraded thresholds and transconductance over time, are controlled by introducing graded or lightly-doped drains (LDD). Using lightly-doped drains increases the source/drain resistance, which adversely effects the gain of the device. If the maximum allowable degradation of I$_{Dsat}$ is 5%, then R$_S$ must be less than 500Ω-μm for a 1μm device with a 200Å gate oxide [2.10]. Series resistance can be reduced by using silicides, but this increases process complexity. If the gates are also silicided (salicide process), the resistivity of the poly can also be reduced. For a silicide process with shallow junctions, care must be taken to prevent depletion of the dopant at the surface, which can lead to Schottky barriers and reduced performance [2.11]. The important considerations in using silicided source/drains are discussed in more detail in chapter 3.

The design of the lightly-doped drain regions entails careful tradeoff between hot electron effects and overlap capacitance. The peak electric field at the drain, which is the source of hot-carrier generation, is highest for very small gate-to-drain overlaps. By designing for a gate-to-drain overlap between 500Å and 1000Å, a good compromise between minimizing overlap capacitance and peak electric field is obtained [2.11]. Overlap capacitance of less than 0.25fF/μm width is desirable.

In addition to overlap considerations, the doping of the LDD regions must be carefully designed to prevent surface-state generation above the N- region which can lead to an additional increase in series resistance and a decrease in transconductance [2.12]. This problem can prevented by using doping levels greater than 5E13 cm^{-3} [2.13].

Parasitic Devices. In designing an overall process flow, care must be taken to ensure that various undesirable parasitic devices do not interfere with device and circuit operation. The parasitics can be separated into two categories: field-effect transistors (FETs) and bipolar transistors [2.14]. Parasitic FETs occur between adjacent source/drains that are separated by a field oxide island. A metal or polysilicon line running over the field oxide between the adjacent diffusions forms the gate. Sufficiently high voltages on the gate electrode may invert the surface underneath the field, causing a leakage path between the diffusions. This leakage leads to voltage shifts on dynamic nodes, DC power dissipation, and noise margin degradation [2.14]. Leakage and punchthrough between source/drain diffusions and adjacent wells are even worse, because the deep well aggravates the short-channel effects. By maintaining adequate spacing between adjacent diffusions and wells (fig. 2.9), and by introducing implants underneath the field oxide to prevent inversion, the parasitic field FETs can be controlled. The field FET implant does increase the capacitance of adjacent junctions. A typical field threshold measured at 1nA is 10 volts or higher.

Parasitic bipolar transistors are primarily a problem in that adjacent NPN and PNP transistors couple to form a parasitic SCR. In a P-epi process, the N+ source/drain, P-epi, and N-well form a lateral NPN, and the P+ source/drain, N-well, and P-epi form a vertical PNP. Sufficient currents can cause a voltage drop inside the well and the substrate to forward bias the emitter-base junctions and turn the transistors on. Because of the positive feedback loop between the

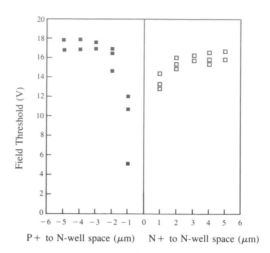

Figure 2.9. Field threshold as a function of source/drain to adjacent well spacing.

two devices, the saturated BJTs act as an SCR in a low impedance high current state known as latchup [2.14]. This can cause temporary or permanent loss of circuit functionality. For permanent latchup to occur, since the peak beta of the PNP will be greater than its ß at the trigger point, the product of the peak dc betas of the two transistors must be greater than 50. This can be prevented by maintaining adequate P+ to N-well spacing (fig. 10). If an epitaxial layer is used as in a high-performance BiCMOS process, the latchup susceptibility is greatly diminished. This effect is discussed in more detail in chapter 4.

2.1.2 Bipolar Devices

As with the MOS transistor, the bipolar transistor can be divided into intrinsic and extrinsic regions. The intrinsic device accounts for the profile design associated with the emitter and the underlying base and collector regions. Extrinsic device concerns include parasitic resistances in series with the base and collector, and capacitances associated with the base-collector and the collector-substrate junctions.

Base Design Issues. For the intrinsic transistor, most of the key tradeoffs come in the base profile design. The base Gummel number Q_b, which is defined as the integral of the base dopant between the two space charge layers of the junctions, sets the collector saturation current and the peak current gain ß. Since ß, which is typically around 100, is inversely proportional to Q_b, the current gain sets a constraint on the maximum base doping allowed. This is traded off against the base resistance of the active base region; an increase in the base doping reduces R_B, which has a favorable impact on the switching speed of

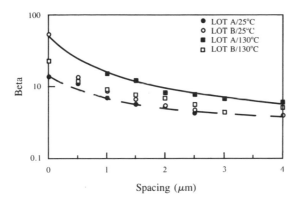

Figure 2.10. Parasitic PNP beta as a function of P+ to N-well spacing. From [2.15]. ©
1987 IEEE.

the device (fig. 2.11). An R_B of 300Ω or less is achievable in high performance
BiCMOS.

The minimum doping level in the base is also constrained by the
requirements of punchthrough and the Early voltage V_A, which is a measure of
the sensitivity of collector current to changes in the base width. Increasing
reverse bias of the collector-base junction causes the spreading of the depletion
region edge into the base, which reduces the neutral base width and increases the
collector current. A similar effect occurs as a function of changes in the bias
across the emitter-base junction. The emitter-base effect is usually only
noticeable for very narrow bases. Doping levels in modern transistors are usually
sufficient so that the Early voltage is not a critical parameter in digital designs.
A value of 15 to 20 volts is acceptable for digital circuits, while analog
applications require an Early voltage of at least 30 volts.

Given the constraints on the base Gummel number, the width of the base
region is minimized to improve the base transit time t_b, which is around 10 to
35ps for a typical base profile. The minimum base width will be constrained by
punchthrough considerations. Punchthrough is suppressed for narrow base
transistors by keeping the peak doping of the base high. In determining the
minimum doping for punchthrough requirements, the statistical fluctuations
associated with the base implant should also be accounted for, since the lowest
dopant concentration volume in the base will determine the punchthrough
voltage [2.16] (fig. 2.12).

The high base doping used to prevent punchthrough also effectively
suppresses high-level injection in the base (Webster effect). High-level injection
occurs in the base when the injection of minority-carrier electrons into the base
becomes comparable to the doping density there. Because of quasi-neutrality

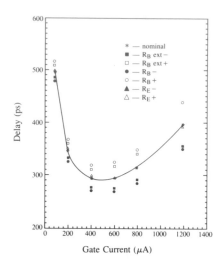

Figure 2.11. Simulated ECL gate delay as a function of current. Nominal parasitic resistances have been varied ± 20%. R_B is the total base resistance, and R_{Bext} is the extrinsic portion.

requirements, a distribution of holes will occur to give a nearly identical profile of holes as electrons in the base [2.18]. To prevent the diffusion of holes in the base and maintain $j_{px}=0$, an opposing electric field is created. Although this electric field aids in the diffusion of the minority carriers across the base, the

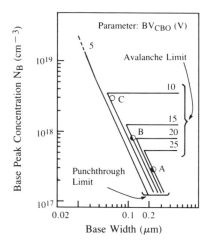

Figure 2.12. Punchthrough and avalanche limits as a function of base impurity concentration and base width (simulated). Statistical doping fluctuations are taken into account. From [2.17]. © 1981 IEEE.

field is obtained at the expense of the available emitter-base junction voltage [2.18]. This reduction in imposed junction voltage is a much more significant effect than the effect of the field-aided diffusion; the net result of high level injection in the base is a decrease in collector current at high emitter biases.

With sufficient high peak doping in the base, the current density is on the order of $1mA/\mu m^2$ before the Webster effect becomes important. The cost of suppressing high-level injection and punchthrough is that the mobility in the base decreases and thus the base transit time suffers. The high doping in the base is beneficial in that the increase in intrinsic carrier concentration caused by bandgap narrowing favorably impacts the ß of the device, as well as decreases the sensitivity of the device to changes in temperature [2.19].

Emitter Design Issues. In conventional bipolar transistor design, the doping of the emitter is made on the order of 1E20 cm^{-3} to improve the emitter efficiency and increase the current gain while keeping R_E low. As the width of the base is decreased to improve performance, the junction depth of the emitter must be decreased to ensure the performance and manufacturability. However, use of conventional metal contacts cause a severe degradation in the current gain. For emitter depths less than $0.25\mu m$, the ß degradation can be a factor of 3 or greater. Polysilicon-contacted emitters can be used to circumvent this problem. Both tunneling [2.20] and a lower mobility in the poly [2.21] have been used to explain the reduced back-injected base current and the resulting increase in gain associated with poly emitters. Which mechanism dominates depends on whether an oxidizing or nonoxidizing clean is used prior to poly deposition [2.22]. In either case, the effect can be modeled by assuming a finite surface recombination velocity at the polysilicon-silicon interface. Typical values are between 3E4 and 7E4 cm/s, where the lower value is achieved using a lower temperature for the emitter anneal. The penalty paid for the increased current gain is an increase in

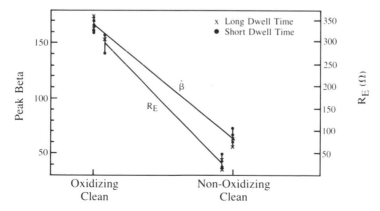

Figure 2.13. Current gain and emitter resistance as a function of polysilicon emitter preclean. Actual emitter size is 1.7 x $6\mu m^2$. From [2.1]. © 1987 IEEE.

emitter resistance, which is minimized and can be controlled by removing the interfacial oxide (fig. 2.13). Emitter resistance on the order of $150\Omega\text{-}\mu m^2$ is typical.

Studies of ECL gate delay as a function of emitter resistance have indicated that R_E is not a *dominant* factor in determining speed for the current level of technology (fig. 2.11). Detailed studies of self-aligned and scaled transistors confirm that even down to emitter widths of $0.5\mu m$, the contribution of all of the R_EC terms is less than 10% [2.23]. This is for light loading conditions. R_E plays a more significant role in BiCMOS gate delay [2.24]. Additionally, for an ECL gate with a very high emitter resistance ($\geq 360\Omega$), the differential voltage gain of an ECL gate falls to the point where the gate will not switch properly.

Either arsenic or phosphorus can been used to dope a polysilicon emitter. Using phosphorus gives an emitter interfacial contact resistance that is a factor of 3 lower and is easier to control. However, to achieve high performance NPNs, shallow emitters and narrow base widths are required. If the phosphorus is used to obtain a shallow emitter, high emitter sheet resistance results. In addition, narrow bases are difficult to achieve with phosphorus because of the emitter push effect. The deeper emitter junction and wider base associated with phosphorus results in an f_T that is 3 times lesser than f_T for an arsenic emitter process [2.25].

An additional design concern for the emitter junction is the hot carriers generated under reverse bias conditions, due to the large electric fields along the periphery of the emitter-base junction [2.26]. The generated carriers tend to increase the fixed oxide charge density, resulting in a degradation in current gain [2.27]. The doping and proximity of the extrinsic base strongly influence the magnitude of this effect. Because of the deeper emitter-base junction and the heavier base doping used, phosphorus poly emitter devices have shown much greater susceptibility to ß degradation than arsenic emitters [2.25].

Collector Design Issues. In the collector, the minimum acceptable doping is set by high level injection considerations (Kirk effect). This effect occurs when the density of electrons transported across the collector-base space charge layer is comparable to the doping in the collector. These carriers cause a spreading out of the electric field and ultimately high injection of both electrons and holes into the collector. This is termed base push-out. The switching speed of the transistor is now degraded due to the increased effective base width and the increased hole storage. The onset of high-level injection occurs at the critical current density J_c given by $J_c = qv_s(N_c + 2\varepsilon_s V_{CB}/qW_c^2)$, where W_c is the effective epi thickness from the base-collector junction to the N+ buried layer [2.47]. Delay in the onset of HLI achieved with increased collector doping (fig.2.14) must be traded off against capacitance C_C and breakdown (BV_{CB0} and BV_{CE0}) considerations. The breakdown voltage requirement will restrict the maximum

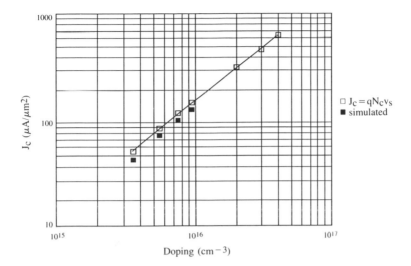

Figure 2.14. Critical current density J_c at which onset of high level injection in the collector occurs, versus N-epi/N-well doping. Effective epi thickness (from the B-C junction to the buried layer) for simulated results is 0.6um.

doping allowed in the collector. The tradeoff between J_c and C_C should be determined according to the operating point of the transistor. For low current densities, the switching speed is limited by charging of the parasitic capacitances, and so C_C should be minimized. For high current density applications, the speed will be limited by the intrinsic device characteristics, and so the collector profile should be optimized to control base push out (fig. 2.15). For high-performance BiCMOS NPNs operated at current densities of

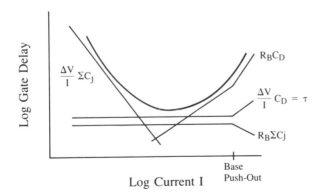

Figure 2.15. At low current, gate delay is determined by the charging of the junction capacitances. At high currents, minority carrier storage associated with high level injection prevails. Curve is composite of effects.

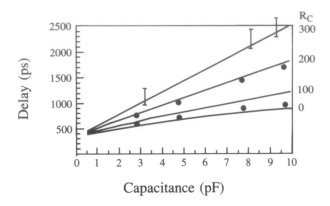

Figure 2.16. BiCMOS gate delay versus load capacitance for various values of R_C. Solid curves: analytical; solid dots: simulation; error bars: experimental. From [2.24]. © 1988 IEEE.

$100\text{-}200\mu A/\mu m^2$, a collector doping of $1E16$ cm^{-3} is typical.

For high performance bipolar transistors, a heavily-doped buried layer is used underneath the epitaxial layer. This is necessary to minimize the collector resistance, which has a dominant effect on the propagation delay of a BiCMOS gate under heavy capacitive loading (fig. 2.16). Like well doping, the thickness of the epi layer will have an impact on high-level injection, collector capacitance, collector resistance, and breakdown. The minimum epi thickness will be set by breakdown voltage (BV_{CE0}), collector capacitance, and

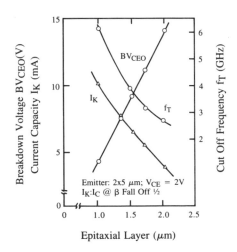

Figure 2.17. Bipolar f_T, maximum current handling capability I_K and BV_{CEO} as functions of epitaxial layer thickness. From [2.28]. © 1987 IEEE.

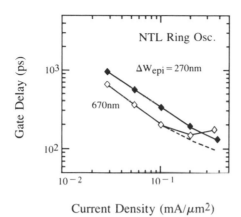

Figure 2.18. Delay-current density characteristics of a non-threshold logic (NTL) ring oscillator gate on thin epi (solid diamond) and thick epi (open diamond) Lines are simulated results. From [2.29]. © 1983 IEEE.

manufacturing controllability considerations. For a BV_{CE0} requirement of 7 volts, it has been reported that the grown epi thickness needs to be greater than approximately 1.25µm (fig. 2.17). However, this will depend on the updiffusion of the buried layer. As in the case of epi doping, C_C and HLI effects are traded off depending upon the operating point of the device (fig. 2.18).

Horizontal Design Issues. Besides vertical constraints, the horizontal dimensions play an important role in the intrinsic device parameters. Since the base resistance is proportional to the width of the emitter and inversely proportional to its length, narrow emitters with W/L aspect ratios of one third or less are preferred. In addition, narrow emitters are superior because emitter debiasing effects are less severe (fig 2.19). This is especially a concern for single-base-contacted devices which are used because of the smaller silicon real estate that they consume compared to dual-base-contacted devices.

As device dimensions approach the one micron level, special considerations should be given to the 2-dimensional dependence of the current gain. This effect has been attributed to the component of the base current which is injected into the sidewall of the emitter. By dividing the bipolar transistor into intrinsic and extrinsic device regions (fig. 2.20), it is seen that the electrical basewidth and the Gummel number for the sidewall-injected carriers is much larger than that of the intrinsic device. As the width of the emitter decreases and the perimeter-to-area ratio increases, the percentage of the overall current gain determined by the sidewall device increases, and the overall current gain is reduced. In decreasing the emitter width from 4.25µm to 0.5µm for a fixed emitter length, a current gain degradation of 65% has been reported due to this effect [2.31].

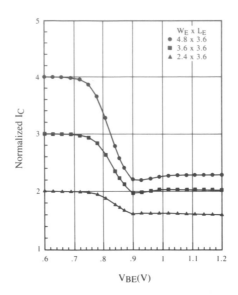

Figure 2.19. DC current delivered by a bipolar transistor, normalized to a W_E x L_E of 1.2 x $3.6\mu m^2$. Devices with wider emitters deliver proportionally less current at higher biases due to emitter debiasing. Results are from quasi-2D simulations.

The impact on device performance can be even more acute as the spacing between the heavily-doped extrinsic base and the emitter edge is decreased. Although reduced spacing decreases R_B, it also increases the effective Gummel number at the edges of the emitter. In the limit, if the lateral diffusion of the extrinsic base extends underneath the emitter, the structure acts like two NPNs in parallel with a wide and narrow base respectively (fig. 2.20). For a $0.5\mu m$ wide device with $0.15\mu m$ encroachment, a degradation in the cutoff frequency of 50% has been reported [2.30]. For a fixed encroachment, a ß degradation of a factor of

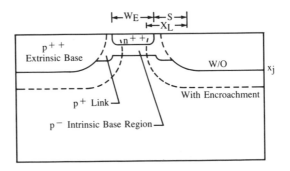

Figure 2.20. Transistor cross section: s is the spacing from extrinsic base implant mask to emitter; x_L is the lateral diffusion from the edge of the implant mask; x_j is the junction depth of the extrinsic base; x_L - s is the lateral encroachment. From [2.30]. © 1987 IEEE.

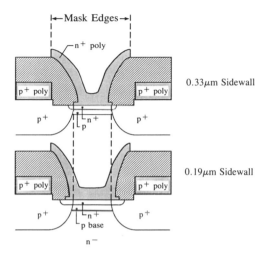

Figure 2.21. Cross section of 2 transistors with the same emitter mask size but different sidewall thickness. The encroachment of extrinsic-base dopants limits the effective area for collector current injection. From [2.32]. © 1987 IEEE.

100 has been seen as the emitter width is reduced from a very wide emitter to a 0.5μm emitter [2.30]. This is due to a decrease in I_C as the heavily-doped extrinsic base under the emitter forms a larger percentage of the total base area under the emitter. Careful design of the base profile at the edges of the emitter combined with shallow emitters seem to minimize these effects.

If the effect of the extrinsic base encroachment is studied as a function of sidewall spacer thickness, then the impact on I_C and I_B can be examined separately. The base current of a device with a shallow emitter is controlled by the surface recombination properties of the poly/monosilicon interface [2.32]; therefore, the base current is proportional to the area of the emitter contact A_c. The collector current, on the other hand, is proportional to the effective area of the emitter A_E, which is confined by the lateral encroachment of the extrinsic base (fig. 2.21). Two effects can result. In the case of negligible encroachment by the extrinsic base on the intrinsic device, the effective emitter area will be larger than the emitter contact area. The ratio of A_E to A_c will increase for decreasing emitter size, and I_C/I_B will increase. If, on the other hand, the extrinsic base does impinge on the active device, then the current gain as a function of geometry will depend on the relationship between the spacer width and the extrinsic base edge. For a fixed extrinsic base edge, if the spacer width is decreased, then the contact area increases while the effective emitter area remains fixed. This translates to an increase in the base current and a reduction in the current gain [2.32]. Therefore, extrinsic base edge location and spacer width can play an important role in the bipolar parameters.

Extrinsic Device Issues. For the extrinsic bipolar device design, the objective is to minimize the parasitic resistances R_B and R_C, and the capacitances C_C and C_S. Since capacitance is proportional to the area and the square root of doping, linear scaling of area is more effective in reducing capacitance than profile modifications [2.19]. Reduced C_S can be achieved by advanced isolation schemes as well as reduced transistor area.

The importance of a low base resistance is clearly seen if ECL gate delay is examined as a function of R_B (fig. 2.11). Generally, the extrinsic portion of R_B is reduced by introducing a heavily-doped region underneath the base contact and extending it toward the emitter. Further reduction in R_B is achieved by minimizing the spacing between the heavily-doped region and the emitter edge. The limit on spacing is set by BV_{EBO}, hot electron degradation considerations, and manufacturing controllability.

In order to achieve low collector resistance, a deep N+ plug is used underneath the collector contact. The need for this will be determined by the epi thickness and doping (or N-well doping) and the operating current density. Without a deep N+ plug, devices operated as low as $70\mu A/\mu m^2$ will have an IR drop in the collector sufficient to forward-bias the collector-base junction (quasi-saturation), which severely degrades performance.

Summary. The impact of the various process parameters on the NPN device characteristics is summarized in Table 2.2. Most of the tradeoffs occur in the design of the base and collector profiles. The coupling between the MOS and NPN characteristics is more clearly revealed in the front-end design section.

Parameter	$\uparrow Q_b$	$\uparrow W_b$	$\uparrow Q_c$
I_C	$\downarrow$	$\downarrow$	-
β	$\downarrow$	$\downarrow$	-
R_B	$\downarrow$	$\downarrow$	$\uparrow$
R_C	-	-	$\downarrow$
I_K	$\uparrow$	$\downarrow$	$\uparrow$
f_T	$\downarrow$	$\downarrow$	$\uparrow$
V_A	$\uparrow$	$\uparrow$	-
V_{PT}	$\uparrow$	$\uparrow$	-
BV_{CBO}	-	-	$\downarrow$
C_E	$\uparrow$	-	-
C_C	-	-	$\uparrow$

Table 2.2. First order effects on NPN parameters, for increasing Q_b, W_b, and Q_c.

Tang and Solomon [2.33] have proposed a method for NPN device design, graphically depicted in fig. 2.22, which summarizes the discussion on NPNs so

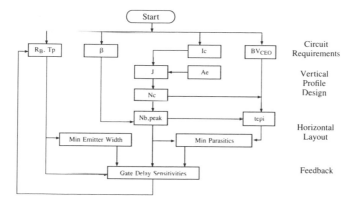

Figure 2.22. First order NPN device design. The vertical profile design is set to optimize device delay, while optimal horizontal design minimizes the parasitics. From [2.34]. © 1988 IEEE.

far. Circuit requirements determine the current density of the device. This sets the minimum collector doping and maximum epi thickness based on Kirk effect considerations. The current density also sets the maximum doping allowed due to the required current gain. Optimization of the base profile is driven by R_B, ß, and base transit time considerations. Narrow emitters and minimization of parasitics lead to an optimized device. If base transit time is the dominant term in the delay equation, performance is not improved by decreasing the horizontal layout.

Required Performance. In designing the devices for a BiCMOS process, the question arises, "How good does an NPN have to be for a given set of MOSFET parameters, and vice versa?" The answer is primarily determined by the circuit requirements. In the case where low-performance NPN transistors are

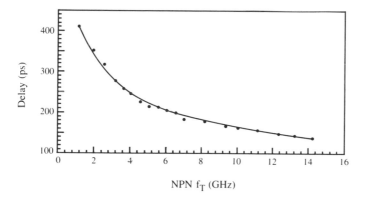

NPN f_T (GHz)

Figure 2.23. BiCMOS gate delay as a function of NPN f_T for a 1μm CMOS technology. From [2.1]. © 1987 IEEE.

used for driving I/Os, then high-performance FETs will be required to drive the substantial capacitance associated with the NPNs. This is achieved by reducing the gate oxide thickness to maximize the FET current drive [2.35]. On the other hand, if high-performance self-aligned bipolar transistors are used, then the need to minimize the gate oxide thickness can be relaxed, provided that the short-channel effects are adequately controlled. Although the use of the high-performance NPN provides a significant improvement in gate delay, improving the intrinsic NPN characteristics beyond a certain level provides diminishing returns in BiCMOS applications (fig. 2.23). Overall, it is best to evaluate the required FET parameters for a given NPN design by evaluating the circuit performance as a function of oxide thickness and channel length [2.35].

2.2 BiCMOS Device Synthesis

In the preceding section, the requirements for good CMOS and bipolar transistors were discussed separately. Quantitative guidelines were given for the device parametrics and the profile characteristics to establish a fundamental understanding of the device requirements.

In this section, the synthesis of BiCMOS will be developed. Because the quantitative foundation has already been given, this section will be more qualitative in scope, with the emphasis on understanding the key compromises imposed in combining bipolar and CMOS on the same chip. Performance compromises can be minimized in the synthesis process, but at the expense of increased process complexity. Most of the compromises in performance are determined by the choices made in the front-end development of the process, where the strongest coupling between device characteristics occurs. A review of these options together with a BiCMOS device synthesis flowchart is discussed below.

2.2.1 Front-end Design

There are a number of different front-end design options available. These include the triple-diffused process, N-epi and P-epi options, and the twin-well process. The resulting performance of the NPNs and MOS transistors, as well as ease of implementation and latchup resistance is compared in Table 2.3. In all cases, front-end design includes the determination of the well profile(s); for the epi options, front-end design additionally includes considerations for the buried layers as well as the epi characteristics.

The triple-diffused option offers ease of implementation and compatibility with CMOS technology. This is obtained at the expense of NPN performance. In particular, triple-diffused NPNs suffer from high R_C, which causes internal

Front-end	Simplicity	CMOS	NPN	Latchup
Triple-diffused	1	1	5	5
P-epi/N-well	3	3	2	2
P-epi/N-well	2	3	4	4
N-epi/Retro P-well	4	5	4	3
π-epi/Twin-well	5	2	2	3

Table 2.3. Ranking for different BiCMOS front-end options, where a 1 is the best. From [2.36]. © 1987 IEEE.

debiasing of the base-collector junction (quasi-saturation) at high currents. R_C can be reduced by increasing the area of the transistor, but the area penalty is usually too high to be practical. In addition, the latchup susceptibility is high because of the high resistivity of the wells and the close proximity of the high-current NPN device. The use of the triple-diffused option may be acceptable in some applications where the primary requirement of the circuit is high performance CMOS [2.36].

The use of an epitaxial layer-based process is required for high-performance BiCMOS. For the CMOS, the epi improves latchup immunity, decreases the impact of the substrate current, and improves the alpha particle immunity in memories [2.36]. The bipolar transistors benefit from reduced collector resistance and suppressed high-level injection in the collector (Kirk effect). Thicker lightly-doped epi is preferred for the CMOS while optimal NPNs are obtained with a thinner, more heavily doped epitaxial layer. To complicate the matter further, parameters chosen for the epitaxial layer are strongly dependent upon well doses, buried layer doses, and well drive time of the process. Thus, the optimization of the epi parameters can be a formidable task. This is discussed in more detail in chapter 3.

2.2.2 BiCMOS Device Compromises

The proposed device design methodology in fig. 2.24 is useful in defining a procedure to evaluate the tradeoffs involved in designing BiCMOS devices. This has been developed using response surface methods, as described in the next section. Although this particular flowsheet is designed to address the key issues for an N-well P-epi process, it is general enough to provide a framework for the subsequent discussion of the major considerations of any epi-based process.

Optimal epitaxial thickness will be a strong function of the particular circuit performance requirements. The bipolar transistor will set a minimum allowed thickness determined by BV_{CEO} constraints (fig. 2.17). Because of the P+ buried layer that is used underneath the NMOS to electrically isolate adjacent N-well tubs, the epi must be thick enough to prevent the updiffusion of this P+ buried layer from excessively increasing NMOS junction capacitance and body effect.

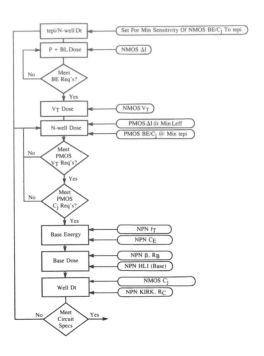

Figure 2.24. BiCMOS device design methodology derived from RSM analysis. After [2.15]. © 1987 IEEE.

This effect will be strongly coupled with the well drive time used. On the other hand, too thick of an epitaxial layer degrades f_T and increases R_C, which may lead to premature quasi-saturation. This possibility is minimized with the use of a deep collector N+ region.

The minimum dose required for the P+ buried layer is set by the N+ to N+ punchthrough, which is a function of the tub-to-tub spacing required for the technology. Coupled with a sufficiently thick epi, keeping the P+ buried layer dose low allows the NMOS junction capacitance and body effect to be decoupled from the P+ buried layer parameters.

The N+ buried layer is usually doped to the solid solubility limit to minimize R_C and latchup immunity. Either arsenic or antimony is used. Depending on the extent of updiffusion and autodoping, the N+ buried layer can play a critical role in the device design compromises.

The required N-well dose is tightly constrained by a number of conflicting needs. To control PMOS short channel effects, the well dose must be sufficient to prevent DIBL. A well doping of 1E16 cm^{-3} or more underneath the base is necessary to suppress base push-out and minimize R_C in the NPN. If the N-well is used to set the PMOS field threshold, then an additional constraint occurs based on the lowest doping that is acceptable at the surface. However, excessive N-well dose is detrimental to PMOS body effect and junction capacitance as well as collector-base capacitance of the NPN.

In the case of the twin-well process, the NMOS transistors will place similar constraints on the allowable P-well dose. An additional degree of freedom is achieved in the P-well design by introducing an additional punchthrough implant below the surface; this effectively decouples DIBL and junction capacitance issues. For the N-well P-epi front-end process, a punchthrough implant is usually required for NMOS channel lengths of ≤1.5μm, depending on the epi concentration used. Since the epi doping must be approximately ten times lower than the N-well doping to prevent degradation of the PMOS parameters (fig. 2.25), twin-well approaches are used with near-intrinsic epi when MOS channel lengths are around 1μm or less.

For the bipolar transistor, the base energy will be constrained by f_T and emitter-base capacitance concerns, while the base dose used is a compromise between current gain and base resistance requirements. Except for the common back-end Dt, the base parameters are determined independently of the CMOS processing requirements.

The dependence of the BiCMOS device parameters on the well Dt is one of the more difficult relationships to quantify because it is intertwined with the choices for epi thickness, buried layer dose, and well dose. Figure 2.26, which results from applying response surface methods to the analysis of the device tradeoffs, shows that the well Dt influences virtually all of the parameters to some degree. The three characteristics that are dominated by the well Dt are the collector resistance and the Kirk effect, restricting the lower end of acceptable Dt, and the NMOS junction capacitance which places an upper bound on the well

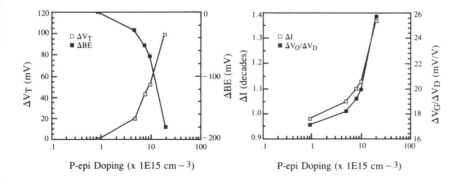

Figure 2.25. For a fixed N-well concentration, maximum P-epi doping is constrained by the effect on threshold voltage (ΔV_T), body effect (ΔBE), and short-channel effects of the PMOS devices.

	Tepi	CSDose	Well Time	Well Dose	V_{T1} Dose	V_{T2} Dose	LMask	Base Dose	Base KeV
NMOS									
$V_T(L)$	2	2	2		D	D			
$V_T(S)$	2	2	2		D	D			
$\Delta V_T(L)$ (BE)	D	D	1		1	D			
$\Delta V_T(S)$ (BE)	D	D	1		1	D	1		
$G_m(S)$					2	2	D		
ΔI	2	2	2		1	D	D		
$G_{DS}(S)$	1	1	2		2	1	D		
C_j	D	1	D		2	2	D		
PMOS									
$V_T(L)$			2	D	D				
$V_T(S)$			2	D	D		2		
$\Delta V_T(L)$ (BE)	1		2	D					
$\Delta V_T(S)$ (BE)	1		2	D			1		
$G_m(S)$				2			D		
ΔI			2	1	1		D		
$G_{DS}(S)$			1	D	1		D		
C_j	2		2	D					
NPN									
BETA								D	1
C_C	2		2	D					1
C_E								1	D
f_T				1				D	D
HLI(C)	1		2	D				1	
HLI(B)								D	1
R_B								D	1
R_C			D	2					2

Figure 2.26. Primary BiCMOS device parameters as a function of key process variables. RSM analysis yields Taylor-series equations, which allow quantification of effects. D=dominant effect, 1=first-order effect, 2=second-order effect.

Dt. The well drive time must be sufficient to completely counterdope the background epi layer. Because the well Dt is typically much smaller for BiCMOS than for a CMOS flow, a significant dip above the N+ buried layer results, leading to undesirable hole storage in the collector of the NPN.

The back-end Dt primarily couples the base width and the MOS source/drain junctions. Back-end Dt must be adequate to achieve good junctions (i.e. low leakage) and relatively low sheet resistance. Excessive back-end Dt leads to increased junction capacitance, increased MOS overlap capacitance, and poorer short channel characteristics. In addition, low back-end Dt is a concern in achieving shallow polysilicon emitters and the resulting narrow base widths. Lower-temperature emitter anneals (900°C) yield higher betas and therefore allow the base resistance to be reduced by increasing the base dose. However, care must be taken to achieve a reproducible poly-single crystal emitter interface. Rapid-thermal anneals may prove to be the vehicle to attain both low R_E and good interface control.

2.3 Design Methodology

2.3.1 Synthesis versus Analysis

In developing an approach for device design, one first must distinguish between the synthesis phase and the analysis phase (fig. 2.27). Synthesis is the

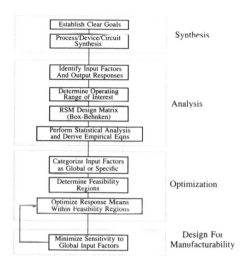

Figure 2.27. Flowchart for efficient application of simulation. From [2.34]. © 1988 IEEE.

development guided by engineering knowledge which leads to a viable concept or process flow. Analysis, on the other hand, is an optimization study which only begins after a viable approach has been conceived. Use of statistical approaches become particularly important at this stage of the development because haphazard approaches can neglect some factors and interactions, and because statistical approaches maximize the efficiency of the studies.

Simulation tools play an important role in both synthesis and analysis phases of development. Process and device simulation tools are integrated with bench test equipment and statistical computer software to provide the appropriate device characteristic curves and parameters (fig. 2.28). Experiments evaluated using simulation tools offer significant savings in both development cost and time compared to similar work done in the pilot line. In addition, the simulation tools provide insight into the device physics which is not normally possible with real devices; this is especially useful during the synthesis stage of development when a viable fabrication concept is being established.

The parameters obtained from simulators must be used cautiously. In particular, semiconductor technology has developed faster than the models used to explain the phenomena. This is particularly true of process models; these become the weakest link in accurate simulation, since parameters from a device simulator are only as accurate as the process profiles that are used. Even in the event that the actual values of the parameters are somewhat inaccurate, the *trends* as a function of a given input parameter are typically correct. In addition, careful comparisons of simulated versus experimental profiles can lead to the

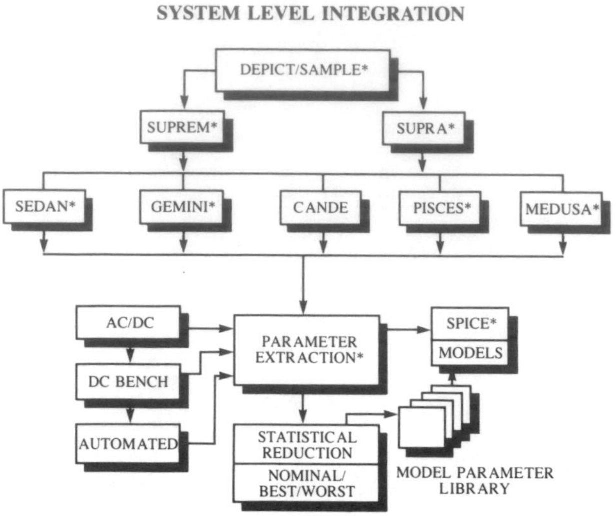

SYSTEM LEVEL INTEGRATION

Figure 2.28. Process and device simulation tools are integrated with parameter extraction tools/routines to provide a statistical model base for designers. After [2.45].

development of more suitable process model coefficients which will give accurate results over a limited range of process perturbations. This method of "fine tuning" is particularly applicable in the analysis stage of development.

After synthesizing a process flow, an experimental design strategy as shown in fig. 2.29 is employed. At first, one's knowledge base is low, so the strategy is to determine from a set of potential factors the key factors having the most influence [2.34]. This can be achieved by using simple statistically-based

DESIGN	SCREENING	FACTORIAL	RESPONSE SURFACE	SPECIAL
Purpose	Identify Important Variables,	Linear & Interaction Effects,	Linear, Interaction & Curvature	Estimate Parameters In
	Crude Prediction Effects	Used for Interpolation	Used For Interpolation	Physical Models
			Used For Optimization	
# Of Factors	6 or More	2–8	2–6	1–4
Tools	Multivari	Fractional Factorials	Central Composite Designs	Special Purpose
	Saturated Designs	Full Factorials	Box-Behnken Designs	
	Variance Breakdown	Split Lot/Nested Designs		
Analysis	Scatter Plots	T&F — Tests	Regression	Regression-
	Multivari Plots	Anova	Anova	Linear/Non-linear
	Data Snooping	Regression	Contour Plots	
	Anova			

Figure 2.29. The Knowledge line as applied to experimentation. Start at the left with screening techniques early in the design when knowledge base is low. From [2.34]. © 1987 IEEE.

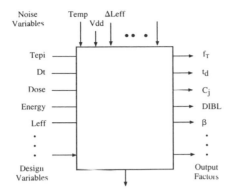

DIBL $= a + b$Tepi $+ c$Dt $+ d$Dose $+ e$Energy $+ f$Leff $+ \dots$

Figure 2.30. The RSM approach is a multiple input-output system treating both design and noise variables, resulting in a second-order Taylor series expression for a given output response. From [2.1]. © 1987 IEEE.

experiments such as Plackett-Burman and factorial designs to screen out unimportant factors.

2.3.2 Response Surface Methodology

Upon determining the key factors for a particular process flow, there is a need to further quantify the region of optimum performance in terms of process variables, as well as evaluate the sensitivities of the device characteristics to manufacturing tolerances. Traditionally, this is done by evaluating the output responses as a function of some set of worse case input parameters. However, this leads to a very limited understanding of the interactions, and does not well quantify the region lying within the space outlined by the worse case corners. In addition, such an approach typically only studies a small subset of the potentially important factors. On the other hand, if factorial designs are used at this stage, the size of the experiments or the number of simulations needed can be overwhelming. For example, a 6-input 3-level full factorial design would require 3^6 or 729 simulations.

Statistical techniques coupled with response surface methods fill the gap between these two extremes. In this approach, the design analysis is viewed as a multiple input-output system that results in a multiple constraints optimization problem. By applying a statistically-based matrix of n trials in N-dimensional space, the response of any output variable can be determined as a function of N independent input factors by a second-order Taylor series equation, which includes all first-order, second-order, and cross product terms [2.34]. This procedure is especially well suited to optimization of a BiCMOS process because of the high number of both key input variables and output responses involved (fig. 2.30).

Motivation. The overall advantage of a statistically-based response surface methodology can be best understood by discussing the results of such a study. 1) The method determines which of the input factors are global (influencing a large number of responses), and which are specific (influencing only a few responses). 2) Since the results give a second-order polynomial model for any response variable as a function of all of the input variables considered, this allows for a complete definition of a response in N-dimensional space; no further simulations are necessary to determine the performance within these input boundaries once these equations are determined. This answers all possible "what if" questions for the defined space. 3) The approach determines the acceptable regions of optimization in which all of the constraints on the output factors are simultaneously satisfied, and thus customer specifications are met. This can be done by using graphical analysis for small problems, or a mathematical grid search for larger problems. In determining acceptable regions of optimization, the approach should be to use specific factors to compensate for the compromises imposed by the global factors. 4) Finally, by examining the shape of the response surfaces, the possible acceptable regions can be ranked in terms of their compatibility with manufacturability. Here, an attempt is made to minimize the sensitivities to process factors which are the most difficult to control. In addition, the process must be evaluated from the viewpoint that it must meet the specifications allowing for the normal process tolerances.

Approach. The steps involved in the application of a statistically-based response surface methodology are outlined below.
1) Choose input factors and response variables using engineering judgment.
2) Choose the low, middle, and high levels of each of the input factors, and determine the desired levels for the response variables.
3) Choose the appropriate experimental design matrix based on statistical guidelines to perform the simulations.
4) Perform the simulations for each response.
5) Use regression analysis to obtain a multi-factor Taylor series approximating each desired response in N-dimensional space.

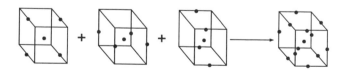

Figure 2.31. Construction of a Box-Behnken design matrix, which is composed of n trials situated on the faces of an N-dimensional cube. From [2.34]. © 1988 IEEE.

6) Analyze the results statistically and from an engineering viewpoint. Use a mathematical grid search, or other techniques to determine the acceptable set of input levels and those with minimal sensitivity to manufacturing variations.
7) Apply the results of the simulations to the design of the process/devices.

A key step is (3), which involves choosing the subset of combinations of inputs that will be simulated. As noted earlier, a full factorial approach leads to large numbers of simulations even for a relatively small number of inputs. Two efficient designs which are applicable to RSM experiments are the central composite design and the Box-Behnken design. The central composite is more appropriate in early development, because it allows for two stage simulation. Box-Behnken designs offer simplicity at the expense of flexibility [2.34]. In a Box-Behnken design, all of the n trials are situated on the faces of an N-dimensional cube (fig. 2.31). This results in a significant simplification of the number of simulations required to develop the Taylor series. Compared to the 729 simulations for a full factorial design, a 6-factor Box-Behnken design requires only 49 simulations. (This excludes the replicates of the center point which are not necessary in simulation modeling.) An example of a 3-factor Box-Behnken design is given in Table 2.4. This significant simplification allows for the evaluation of a relatively large number of input variables in a short period of time.

X1	X2	X3
+1	+1	0
+1	-1	0
-1	+1	0
-1	-1	0
+1	0	+1
+1	0	-1
-1	0	+1
-1	0	-1
0	+1	+1
0	+1	-1
0	-1	+1
0	-1	-1
0	0	0

Table 2.4. Three-variable Box-Behnken design. Each factor represents an experimental point, where +1 is the high, 0 is the nominal, and -1 is the low case of a 3-level design. From [2.34]. © 1988 IEEE.

Analysis. After the simulations are run, the results are compiled, and a statistical analysis package (such as SAS - Statistical Analysis System) is used to determine the least-squares coefficients associated with the Taylor series

$C_j = 2.23 - 0.185*TEPI + 0.162*WTIME + 0.112*CSDOSE$

$G_mL = 7.42 - 0.025*WTIME - 0.179*VT1 - 0.138*VT2 - 0.016*TEPI^2$

$G_mS = 289 - 7.0*VT1 - 6.92*VT2 - 46.6*LMASK + 8.06*LMASK^2$

$V_TL = 680 + 46.7*VT1 + 47.0*VT2$

$V_TS = 613 + 47.4*VT1 + 11.4*LMASK + 51.9*VT2$

$BaseRho = 3226 - 1252*BD - 608*BE + 293*BD*BE + 487*BD^2 + 239*BE^2$

$Beta = 47.2 - 31.6*BD - 2.5*BD*BE + 15.6*BD^2$

$C_C = 4.00 - 0.018*TEPI + 0.317*WDOSE - 0.031*WTIME - 0.116*BE$

$C_E = 2.44 + 0.362*BD - 0.713*BE - 0.168*BD*BE + 0.204*BE^2$

$f_T = 4.86 - 0.568*BD - 1.76*BE - 0.75*BD*BE$

Figure 2.32. Subset of reduced equations for BiCMOS process. WTIME is the well drive time, TEPI is the epi thickness, CSDOSE is the P+ buried layer dose, BD is the base implant dose, BE is the base implant energy, VT1 is the threshold adjust implant dose, VT2 is the punchthrough implant dose, LMASK is the mask length, L is the long-channel device, S is the short-channel device.

model. The resulting multi-factor second-order Taylor series equations are lengthy and therefore cumbersome to apply mathematical grid search techniques to, as well as difficult to analyze qualitatively. By using Analysis of Variance (ANOVA), the terms which are relatively unimportant to the response can be eliminated. The adequacy of the reduced model can be verified by examination of residuals and R^2. A set of simplified equations for a P-epi N-well BiCMOS process is given in fig. 2.32.

Such a set of equations provides insight into interactions and tradeoffs which are not readily foreseen in conventional approaches. For example, a table can be generated which shows which inputs affect each of the responses. The sparsity of this matrix is used to separate the inputs into global and specific factors. Further insight can be obtained by examining the relative magnitude of each of the coefficients within each response equation, and then ranking the importance of the inputs for each of the responses, as shown in fig. 2.26. From the table, it is seen that epitaxial thickness and well time are global, well dose and mask length (poly width) are quasi-global, and the remaining factors are specific. By categorizing the variables as such, it is easier to see which specific factors need to be adjusted for the compromises resulting from the global factors.

In addition, two-dimensional contour plots can be obtained, which provide a qualitative feel for the process tradeoffs. An example of such is given in fig. 2.33. Combining the table of key responses (fig. 2.26) with the contour plots leads to the development of a methodology for analyzing the key tradeoffs, which was given in fig. 2.24. Without the use of response surface methods, it would be difficult to get a handle on all of the tradeoffs involved in a BiCMOS process.

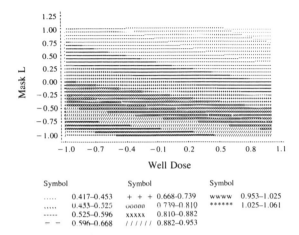

Symbol		Symbol		Symbol	
.....	0.417–0.453	+ + +	0.668–0.739	wwww	0.953–1.025
.....	0.453–0.525	ooooo	0.739–0.810	******	1.025–1.061
-----	0.525–0.596	xxxxx	0.810–0.882		
– –	0.596–0.668	/ / / / /	0.882–0.953		

Figure 2.33. Two-dimensional response surface of PMOS ΔI as a function of mask length and well dose. From Taylor-series equations of RSM analysis.

In addition to using the graphical method for qualitative analysis, a mathematical grid search technique can be used to determine the set of possible input parameters that will satisfy a given set of desire response variables. To accomplish this, a grid of points must be derived for each input factor. Then all of the possible combinations of grid points for the input factors are searched for those which yield all responses in their required ranges simultaneously. Further optimization can be achieved by defining targets for each parameter and using desirability functions to determine the best set of inputs based on these targets. By evaluating the perturbations in the output responses as a function of small perturbations in the input variables (i.e. manufacturing variations), desirability expected loss functions can be applied to both optimize the response to a given target and to minimize the responses to the influence of manufacturing variations [2.45].

2.3.3 Circuit Design Considerations

The above scenario provides a set of optimal process inputs (i.e. implant doses and energies, drive times, etc.) based on targets and manufacturability considerations. However, in the early stages of product development, device targets may not be well defined. Ultimately, the determination of these targets should be strongly coupled to the requirements of a particular circuit design. The design requirements add an additional level of hierarchy to the optimization of device parameters. The design requirements must be met not only at the nominal operating conditions, but also over the range of temperature and power supply conditions.

A methodology has been proposed for the evaluation of BiCMOS worse-case (corners) model parameters [2.37]. The key is to develop a self-consistent set of NPN and MOSFET parameters that are physically meaningful. This can be achieved by using numerical simulation tools coupled with the insights provided by analytical expressions.

NPN model parameters are separated into those dominated by the base doping Q_b and the collector doping Q_c (Table 2.2) to establish the appropriate NPN corners. A third dimension added to the matrix is lithographic variation, which for the example of a non-self-aligned NPN may have a more significant impact on the base resistance than the variation in the base sheet resistance. Expected process variations in Q_b and Q_c are then used to determine the expected variation in the design model parameters through the use of appropriate analytical expressions or simulations [2.37]. In optimizing the NPN, the primary tradeoff is in the base transit time which determines f_T versus the base resistance which impacts the gate delay.

For the case of the MOSFETs, the primary parameters of variability are the channel width, channel length, gate oxide capacitance, and the threshold voltage [2.37]. By choosing an appropriate set of key parameters and limiting the number of corners examined, most of the variation in MOS behavior can be accounted for in a compact manner.

Because the variation in MOSFET parameters has a stronger impact on circuit performance than the variability due to the NPN parameters, the BiCMOS corners are based on the FET corners with the NPNs being chosen according to these. The physical correctness of the parameter sets must be guaranteed by examining the device model parameter correlation to the process parameters for each selected corner [2.37]. The overall procedure allows for the evaluation of design performance as a function of model parameter variability and temperature conditions. The results of such an analysis clearly show the superior robustness of BiCMOS compared to CMOS (fig. 2.34).

2.4 Device Scaling

2.4.1 MOS Scaling

Scaling is used both to increase the functional density and the speed of the circuits. However, if horizontal dimensions are scaled by the factor K ($K < 1$) without any modifications to the vertical profile, the close proximity of the source and drain junctions leads to DIBL and the associated short channel effects. This can be circumvented by applying the classical scaling theory [2.38] to maintain constant shape and magnitude of the electric field in the device. By scaling the voltages (including the threshold voltage) by K and the doping by $1/K$, the depletion widths are essentially scaled, and the electric fields in a

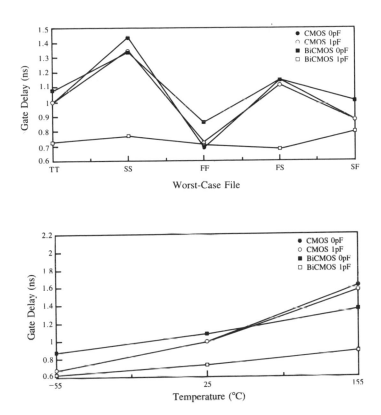

Figure 2.34. a) Corners circuit simulation based on physically-based worse case process files. FF=Fast-Fast, SS=Slow-Slow, TT=Typical, where the 2 identifiers refer to the PMOS and NMOS parameters respectively. The robustness of the BiCMOS under loaded conditions is clearly seen. b) BiCMOS and CMOS normalized gate delay as a function of temperature for unloaded and loaded conditions from TT process file. From [2.37]. © 1988 IEEE.

short-channel device looks the same as those in a long-channel device. Thus long-channel behavior is maintained.

Unfortunately, in such a scenario, the device is harder to turn off because the subthreshold swing S does not scale [2.3]. In addition, off-chip drive requirements have placed a constraint on supply voltage. Therefore an alternate scaling scheme is to scale horizontal dimensions and dopings as before, but fix the voltages. The gate oxide can be scaled less aggressively to minimize the increase in electric field and the reduction in mobility [2.39]. However, the severity of the electric field will eventually make this scaling method unacceptable.

A proposed generalized scaling theory is summarized in Table 2.5. Physical dimensions are scaled by K and potentials are scaled by a smaller factor ß. Local fields are allowed to increase by K/ß, but the shape of the electric field is conserved. By scaling voltages less aggressively, two concerns are addressed [2.40]: 1) V_T should not be scaled as aggressively as horizontal dimensions, since V_T varies with channel length, drain voltage, process variability, and temperature. 2) The junction built-in potential does not scale, and therefore the

Parameter	Scale Factor
Linear dimensions	K
Gate oxide	K
Junction depth	K
Voltage	ß
Channel doping	$ß/K^2$
Bulk doping	$<ß/K^2$
Electric field	$ß/K$
Capacitance	$1/K$
Current	$ß^2/K$
Current density	$ß^2/K^3$
Power density	$ß^3/K^3$
Delay	$K^2/ß$
Time constant	1

Table 2.5. Generalized MOSFET scaling rules, where K < 1. From [2.40]. © 1984 IEEE.

depletion width becomes a larger percentage of the channel length for small devices.

Instead of strictly adhering to the general scaling rules, modifications are made to improve device characteristics. By scaling the bulk doping less than the channel doping [2.1], capacitance and substrate sensitivity to bias (body effect) are minimized while maintaining control of the short-channel effects. Also, by scaling junction potential less aggressively than K, source/drain resistance can be improved without sacrificing control of drain-induced barrier lowering.

2.4.2 Bipolar Scaling

The threshold voltage of the bipolar device, or the voltage necessary to turn it on, is determined by the bandgap of the material. Therefore, although threshold does not show the same sensitivity to geometry and process parameters as the MOS transistor, it also is not a parameter that can be reduced. This means that the power supply voltages will generally not be scaled.

To maintain peak performance in scaling, device dimensions should be scaled by K [2.41]; this implies that the current density will increase by $1/K^2$. To minimize the base transit time, base width is scaled by $K^{0.8}$. To prevent punchthrough and tunneling, the base doping is scaled by the inverse of the base width squared. This leads to a reduction in the base resistance and the $R_B C_C$ time constant. Although the mobility of the minority carriers is degraded in the base, this is offset by the base bandgap narrowing, which improves the current gain.

Optimization of device performance will depend on the current density at which the device is operated. At high current densities, optimization of the intrinsic device is important. It follows that base transit time should be minimized and the Kirk effect alleviated by scaling the collector doping in proportion to the current density. For low current densities, minimization of parasitic capacitances is critical. This can be achieved by going to self-aligned schemes and trench isolation. Polysilicon emitters are used to obtain the required base width and still maintain a high emitter efficiency. So, unlike the MOS transistor, where a cross-section of a scaled device looks basically the same as a "conventional" device of 10 years ago, the high-performance scaled bipolar device is radically different from a conventional one.

The bipolar device scaling scheme is summarized in Table 2.6. In addition to the device requirements, optimum circuit performance requires that the capacitor and resistor ratios should be held constant [2.42]. This optimizes the performance for power-delay considerations.

Additional concerns may require that some of the scaling rules be relaxed. Particular concerns include emitter resistance and current density at the emitter contact [2.42]. In addition, if all of the interconnect dimensions are scaled by K, the current density increases by $1/K^2$, which has serious implications for electromigration. By increasing the width of the power bus, the problem is alleviated. In general, VLSI circuits become metal limited because of the space required by the metal lines and the degradation in speed associated with the R_C

Parameter	Scale Factor
Device dimensions	K
Voltage	~1
Base width	$K^{0.8}$
Base doping	$1/W^2$
Collector doping	J
Delay	K
Current density	$1/K^2$
Depletion capacitance	K

Table 2.6. BJT scaling, where K < 1, W is the base width, and J is the current density. After [2.33].

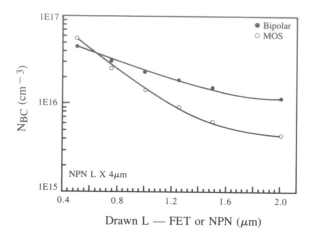

Figure 2.35. The near surface doping requirements of the MOS and the bulk doping requirements of the NPN converge as MOS and bipolar devices are scaled. From [2.36]. © 1987 IEEE.

delay [2.43].

2.4.3 BiCMOS Scaling Considerations

As the technology is scaled to the submicron level, the MOSFETs require increasing subsurface doping to control the short-channel effects, and the bipolar transistors need increasing well doping to delay base push-out (Kirk effect). Therefore, the profile requirements for the two devices converge as the devices are scaled (fig. 2.35). This relaxes some of the device tradeoffs between the two transistor types.

The result of scaling power supply voltage can have a significant impact on BiCMOS circuit performance, depending on the circuit configuration (fig. 2.36). Although BiCMOS offers a speed advantage for driving capacitive loads with a 5V supply, the gate delay for a BiCMOS gate that does not swing from rail to rail degrades much more rapidly as supply is decreased; it loses its advantage altogether   for light capacitive loads ($\leq$ 1pF) when V_{DD} is less than 3 volts. This is not the case for some circuits, as discussed in chapter 5. An additional concern is that for ECL subcircuits built on a BiCMOS chip using this low supply level, 3-level series gating is lost [2.1]. Therefore, the implications of scaling BiCMOS below 0.5µm must be evaluated carefully to achieve optimum performance. One way to take advantage of BiCMOS in a scaled technology is to build an on-chip voltage regulator to run the CMOS at a lower voltage while using standard TTL or ECL voltage levels for the bipolar I/Os.

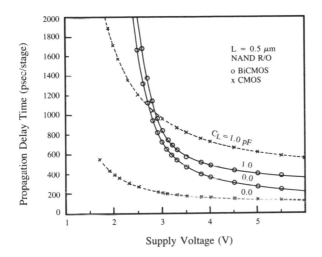

Figure 2.36. BiCMOS and CMOS gate delay measured from NAND ring oscillators as a function of supply voltage. From [2.44]. © 1987 IEEE.

References

2.1] A. R. Alvarez, "BiCMOS Technology," 1987 IEDM Short Course on BiCMOS Technol., Dec. 1987.

2.2] A. G. Sabnis and J. T. Clemens, "Characterization of the Electron μ in the Inverted <100> Si Surface," 1979 IEDM Tech. Dig., pp. 18-21, Dec. 1979.

2.3] Y. P. Tsividis, *Operation and Modeling of the MOS Transistor*, Mc Graw-Hill, New York, 1987.

2.4] R. R. Troutman, "VLSI Limitations from Drain-Induced Barrier Lowering," IEEE J. Solid-State Circuits, Vol. SC-14, No. 4, pp. 383-391, April 1979.

2.5] C. Hu, S. C. Tam, F.-C. Hsu, P.-K. Ko, T.-Y. Chan, K. W. Terrill, "Hot-Electron-Induced MOSFET Degradation - Model, Monitor, and Improvement," IEEE J. Solid-State Circuits, Vol. SC-20, No. 1, pp. 295-305, Feb. 1985.

2.6] J. R. Brews, "Physics of the MOS Transistor," in *Applied Solid-State Science*, Academic Press, New York, 1981.

2.7] K. M. Cham, S.-Y. Chiang, "Device Design for the Submicrometer P-Channel FET with N+ Polysilicon Gate," IEEE Trans. Electron Devices, Vol. ED-31, No. 7, pp. 964-968, July 1984.

2.8] G. J. Hu and R. H. Bruce, "Design Tradeoffs between Surface and Buried-Channel FETs," IEEE Trans. Electron Devices, Vol. ED-32, No. 3, pp. 584-588, March 1985.

2.9] S. J. Hillenius and W. T. Lynch, "Gate Material Work Function Considerations for 0.5 Micron CMOS," Proceedings, IEEE Int'l Conf. on Computer Design - VLSI in Computers, pp. 147-150, Oct. 1985.

2.10] G. Sh. Gildenblat and S. S. Cohen, "Criteria for Estimating the Impact of Series Resistance on MOSFET Performance," Solid-State Electronics, Vol. 31, No. 2, pp.

261-263, Feb. 1988.

2.11] R. Haken, R. Chapman, T. Holloway, C.-F. Wan, D. Bell, B. Gale, and T. Tang, "Transistor Source/Drain LDD Design Issues and Tradeoffs for Submicron CMOS," Proceedings, IEEE Int'l Conf. on Computer Design - VLSI in Computers, pp. 151-154, Oct. 1985.

2.12] F.-C. Hsu and S. Tam, "Relationship between MOSFET Degradation and Hot-Electron-Induced Interface-State Generation," IEEE Electron Device Lett., Vol. EDL-5, No. 2, pp. 50-52, Feb. 1984.

2.13] M. Kinugawa, M. Kakumu, S. Yokogawa, K. Hashimoto, "Submicron MLDD NMOSFETs for 5V Operation," 1985 Dig. of Technical Papers, Symposium on VLSI Technology, pp. 116-117, May 1985.

2.14] J. Y. Chen and A. G. Lewis, "Parasitic Transistor Effects in CMOS VLSI," IEEE Cir. and Devices Mag., Vol. 4, No. 3, pp. 8-13, May 1988.

2.15] A. R. Alvarez, J. Teplik, D. W. Schucker, T. Hulseweh, H. B. Liang, M. Dydyk, I. Rahim, "Second Generation Bi-CMOS Gate Array Technology," 1987 IEEE Bipolar Circuits and Technol. Meeting, pp. 113-117, Sept. 1987.

2.16] D. K. Ferry, L. A. Akers, E. W. Greeneich, *Ultra Large Scale Integrated Microelectronics*, Prentice Hall, Englewood Cliffs, N. J., 1988.

2.17] N. Hanaoka and A. Anzai, "Perspective of Scaled Bipolar Devices," 1981 IEDM Tech. Dig., pp. 512-515, Dec. 1981.

2.18] R. M. Warner and B. L. Grung, *Transistors, Fundamentals for the Integrated-Circuit Engineer*, John Wiley and Sons, New York, 1983.

2.19] D. D. Tang, P. M. Solomon, T. H. Ning, R. D. Isaac, R. E. Burger, "1.25μm Deep-Groove-Isolated Self-Aligned Bipolar Circuits," IEEE J. Solid-State Circuits, Vol. SC-17, No. 5, pp. 925-931 , Oct. 1982.

2.20] H. C. De Graaff, J. G. De Groot, "The SIS Tunnel Emitter: a Theory for Emitters with Thin Interface Layers," IEEE Trans. Electron Devices, Vol. ED-26, No. 11, pp. 1771-1776, Nov. 1979.

2.21] T. K. Ning, R. D. Isaac, "Effect of Emitter Contact on Current Gain of Silicon Bipolar Devices," IEEE Trans. Electron Devices, Vol. ED-27, No. 11, pp. 2051-2055, Nov. 1980.

2.22] P. Ashburn and B. Soerowirdjo, "Comparison of Experimental and Theoretical Results on Polysilicon Emitter Bipolar Transistors," IEEE Trans. Electron Devices, Vol. ED-31, No. 7, pp. 853-860, July 1984.

2.23] E.-F. Chor, A. Brunnschweiler, P. Ashburn, "A Propagation-Delay Expression and its Application to the Optimization of Polysilicon Emitter ECL Processes," IEEE J. Solid-State Circuits, Vol. SC-23, No. 1, pp. 251-259, Feb. 1988.

2.24] E. W. Greeneich and K. L. McLaughlin, "Analysis and Characterization of BiCMOS for High-Speed Digital Logic," IEEE J. Solid-State Circuits, Vol. SC-23, No. 2, pp. 558-565, April 1988.

2.25] B. Landau, B. Bastani, D. Haueisen, R. Lahri, S. Joshi, J. Small, "Poly Emitter Bipolar Transistor Optimization for an Advanced BiCMOS Technology," IEEE 1988 Bipolar Circuits and Technology Meeting, pp. 117-120, Sept. 1988.

2.26] D. Burnett and C. Hu, "Hot-Carrier Effects in Polysilicon Emitter Bipolar Transistors," IEEE 1988 Bipolar Circuits and Technology Meeting, pp. 95-98, Sept. 1988.

2.27] S. P. Joshi, R. Lahri, and C. Lage, "Poly Emitter Bipolar Hot Carrier Effects in an Advanced BiCMOS Technology," 1987 IEDM Tech. Dig., pp. 182-185, Dec. 1987.

2.28] T. Ikeda, A. Watanabe, Y. Nishio, I. Masuda, N. Tamba, M. Odaka, K. Ogiue, "High-Speed BiCMOS Technology with a Buried-Layer Twin Well Structure," IEEE Trans. Electron Devices, Vol. ED-34, No. 6, pp. 1304-1310, June 1987.

2.29] D. D. Tang, K. P. Mac Williams, P. M. Solomon, "Effects of Collector Epitaxial Layer on the Switching Speed of High-Performance Bipolar Transistors," IEEE Electron Device Lett., Vol. EDL-4, No. 1, pp. 17-19, Jan. 1983.

2.30] D. D. Tang, T.-C. Chen, C.-T. Chuang, G. P. Li, J. Stork, M. B. Ketchen, E. Hackbarth, T. H. Ning, "Design Considerations of High-Performance Narrow-Emitter Bipolar Transistors," IEEE Electron Device Lett. , Vol. EDL-8, No. 4, pp. 174-175, April 1987.

2.31] D. P. Verret and J. E. Brighton, "Two-Dimensional Effects in the Bipolar Polysilicon Self-Aligned Transistor," IEEE Trans. Electron Devices, Vol. ED-34, No. , pp. 2297-2303, Nov. 1987.

2.32] P.-F. Lu, G. P. Li, D. D. Tang, "Lateral Encroachment of Extrinsic Base Dopant in Submicrometer Bipolar Transistors," IEEE Electron Device Lett. , Vol. EDL-8, No. 10, pp. 496-498, Oct. 1987.

2.33] P. M. Solomon and D. D. Tang, "Bipolar Circuit Scaling," 1979 IEEE Int'l Solid-State Circuits Conf., pp. 86-87, Feb. 1979.

2.34] A. R. Alvarez, B. L. Abdi, D. L. Young, H. D. Weed, J. Teplik, and E. R. Herald, "Application of Statistical Design and Response Surface Methods to Computer-Aided VLSI Device Design," IEEE Trans. Computer-Aided Design, Vol. CAD-7, No. 2, pp. 272-288, Feb. 1988.

2.35] A. R. Alvarez, D. W. Schucker, "Bi-CMOS Technology for Semi-Custom Integrated Circuits," IEEE Custom Integrated Circuits Conference, pp. 22.1.1-22.1.5, 1988.

2.36] A. R. Alvarez, J. Teplik, H. B. Liang, T. Hulseweh, D. W. Schucker, K. L. Mc Laughlin, K. A. Hansen, B. Smith, "VLSI BiCMOS Technology and Applications," 1987 Int'l Symposium on VLSI Technology, Systems, and Applications, pp. 314-319, March 1987.

2.37] A. R. Alvarez, J. Arreola, S. Y. Pai, and K. N. Ratnakumar, "A Methodology for Worse-Case Design of BiCMOS Integrated Circuits," 1988 IEEE Bipolar Circuits and Technology Meeting, pp. 172-175, Sept. 1988.

2.38] R. H. Dennard, F. H. Gaensslen, H. N. Yu, V. L. Rideout, E. Bassons, and A. R. LeBlanc, "Design of Ion-Implanted MOSFETs with Very Small Physical Dimensions," IEEE J. Solid-State Circuits, Vol. SC-9, No. 5, pp. 256-268, Oct. 1974.

2.39] P. K. Chatterjee, W. R. Hunter, T. C. Holloway, Y. T. Lim, "The Impact of Scaling Laws on the Choice of N-Channel or P-Channel for MOS VLSI," IEEE Electron Device Lett., Vol. EDL-1, No. 10, pp. 220-223, Oct. 1980.

2.40] G. Baccarani, M. R. Wordeman, and R. H. Dennard, "Generalized Scaling Theory and its Application to a 1/4 Micrometer MOSFET Design," IEEE Trans. Electron Devices, Vol. ED-31, No. 8, pp. 452-462, April 1984.

2.41] D. D. Tang, P. M. Solomon, "Bipolar Transistor Design for Optimized Power-Delay Logic Circuits," IEEE J. Solid State Circuits, Vol. SC-14, No. 4, pp. 679-684, Aug. 1979.

2.42] T. H. Ning, D. D. Tang, and P. M. Solomon, "Scaling Properties of Bipolar Devices," 1980 IEDM Tech. Dig., pp. 61-64, Dec. 1980.

2.43] J. S. Huang, "Bipolar Technology Potential for VLSI," VLSI Design, Vol. 4, pp.

64-66, July 1983.

2.44] H. Momose, K. M. Cham, C. I. Drowley, H. R. Grinolds, H. S. Fu, "0.5 Micron BiCMOS Technology," 1987 IEDM Tech. Dig., pp. 838-840, Dec. 1987.

2.45] D. L. Young, J. Teplik, H. D. Weed, N. Tracht, A. R. Alvarez, "Application of Statistical Design and Response Surface Methods to Computer-Aided VLSI Device Design II: Desirability Functions and Taguchi Methods," submitted to IEEE Trans. Computer-Aided Design.

2.46] A.R. Alvarez, P. Meller, D. Schucker, F. Omerod, J. Teplik, B. Tien, D. Maracas, "Technology Considerations in Bi-CMOS Integrated Circuits," Proceedings, IEEE Int'l Conf. on Computer Design - VLSI in Computers, pp. 159-163, Oct. 1985.

2.47] S. K. Ghandi, *Semiconductor Power Devices*, Wiley, New York, 1977.

Chapter 3

BiCMOS Process Technology

R.A. Haken, R.H. Havemann, R.H. Eklund and L.N. Hutter
(Texas Instruments, Inc.)

3.0 Introduction

For high performance LSI and VLSI digital circuit applications, BiCMOS technology has become predominantly driven from a CMOS processing base. The principle reason for this is that LSI and VLSI digital BiCMOS circuits tend to be CMOS-intensive because of power dissipation limitations (for example, high density ECL I/O SRAMs and gate arrays). The CMOS-intensive nature of these circuits requires a process technology that will result in the highest possible CMOS performance. Consequently, BiCMOS fabrication technology tends to be CMOS-based, and the process steps needed to realize a high performance bipolar device are usually merged with a core CMOS process flow [3.1, 3.2, 3.3]. In the case of analog BiCMOS, the increasing demand to have on-board digital logic integration has also resulted in these processes being CMOS-oriented.

As CMOS technology has been extended to cross the "1 μm discontinuity" into the submicron regime, the structural requirements for realizing high performance CMOS and bipolar transistors have tended to converge. For example, it is common practice in submicron CMOS processes to incorporate silicided gates and diffusions to achieve lower sheet resistance (from typically 20-50 ohms/sq to 1.5 ohms/sq). Figure 3.1 shows how the same silicidation technology can be used to simultaneously enhance the performance of both CMOS devices and polysilicon emitter bipolar transistors. For the bipolar device, the silicidation process can be used to minimize the emitter resistance by cladding the polysilicon in a similar manner as is done for the CMOS gates. It can also be used to decrease the extrinsic base resistance by cladding the P+ diffusion, again as is done for the PMOS source/drains.

Many other process steps in a submicron CMOS and bipolar process can be merged and shared to realize a high performance BiCMOS technology with the minimum of added complexity, compared to a core CMOS process flow. One example of how a key process feature of one device can be advantageously shared with the other is the case of a buried N+ layer. Buried N+ layers are common features in bipolar processes and are introduced to minimize collector resistance. However, the buried layer can also be utilized in CMOS circuits to reduce latchup susceptibility if it is placed underneath the N-wells containing PMOS devices, as shown in Fig. 3.1. This results in the ultimate "retrograde" CMOS well profile and eliminates the need for epitaxial starting material, which is often used in submicron

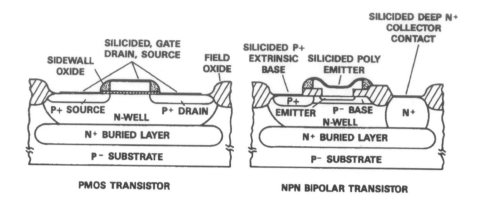

Figure 3.1 Illustration of the trend in submicron BiCMOS towards merged CMOS-bipolar structures and shared process steps. In this example the polysilicon gate CMOS and polysilicon emitter bipolar devices have been simultaneously silicided to reduce gate, emitter, source/drain, extrinsic base and deep N+ collector sheet resistances to 1.5 ohms/sq.

CMOS to prevent latchup [3.4, 3.5].

By judicously merging process steps that are necessary for the individual fabrication of bipolar and CMOS devices, high performance submicron BiCMOS technologies can be realized with only an additional 3-4 masking levels, compared to a baseline submicron CMOS technology. Since, for a number of applications, the same generation BiCMOS can realize up to a 2X performance improvement over CMOS alone, for a cost increase of only 1.3-1.5X, the attractiveness of BiCMOS becomes obvious. This is particularly the case when one considers that the same manufacturing capability can be used. For CMOS alone to realize such a performance improvement, it would be necessary to implement at least the next generation technology, and the high cost of capital equipment makes this approach a much less cost-effective method of realizing higher performance.

In the following sections of this chapter, the process requirements and techniques for realizing a high performance BiCMOS technology are reviewed. Throughout these sections, the merging and sharing of processes to simultaneously satisfy the bipolar and CMOS requirements while reducing process complexity are highlighted. Evolution of BiCMOS from a CMOS processing perspective is discussed, showing how the simplest to highest performance BiCMOS processes are realized. Bipolar and CMOS isolation considerations are reviewed and the tradeoffs involved in chosing different types of bipolar and CMOS device structures are discussed. This includes the various approaches for bipolar emitter, base and collector processing and compatibility with CMOS. The relationship between CMOS N-well and bipolar collector design is also reviewed in detail. A subsequent section gives a comprehensive overview of interconnect processes applicable to BiCMOS, such as silicidation, local interconnect, planarization and contact and via "plugging". A typical submicron BiCMOS process flow is presented in a later section. The process flow is described in cross-sectional form so as to allow the reader to comprehend how the various processing levels are combined to realize the isolation, bipolar and CMOS structures and interconnects. In the final section of the chapter, considerations for analog BiCMOS process design are reviewed together with a typical process flow.

3.1 Evolution of BiCMOS From a CMOS Perspective

For the majority of LSI and VLSI BiCMOS circuits, such as SRAM's and gate arrays, the designs tend to be CMOS-intensive because of power dissipation limitations associated with the bipolar circuitry. Consequently, BiCMOS technologies have tended to evolve from CMOS processes in order to obtain the highest CMOS performance possible. The bipolar processing steps have then been added to the core CMOS flow to realize the desired device characteristics. The evolution of BiCMOS from a CMOS perspective is reviewed below, from the simplest implementation to the highest-performance structures.

Figure 3.2 shows a cross-section of a basic N-well CMOS structure, typical of what might be fabricated with 3 μm design rules. The NMOS device, built in a 15 μm thick P- epitaxial layer on top of a P+ substrate, is shown on the lefthand side of the cross-section while the PMOS transistor, built in an implanted N-well approximately 5 μm deep, is shown on the right. The P+ substrate is used to reduce latchup susceptibility by providing a low impedance path through a vertical PNP device. Polysilicon gates are used for both the PMOS and NMOS transistors, which typically have a 40 nm gate oxide thickness and a 0.6 μm N+ and P+ source/drain junction depth.

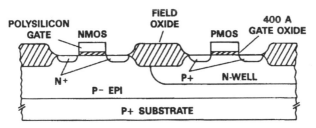

Figure 3.2 Basic N-well CMOS process cross-section.

The simplest way in which an NPN bipolar transistor can be added to this basic CMOS structure is by using the PMOS N-well as the collector of the bipolar device [3.6] and introducing an additional mask level for the P-base region (Fig. 3.3). The P-base is approximately 1 μm deep with a doping level of about 1×10^{17} atoms/cm³. The N+ source/drain ion implantation step is used for the emitter and collector contact of the bipolar structure. The P+ source/drain ion implantation step is used to create a P+ base contact to minimize the base series resistance. This approach results in a BiCMOS technology where the following process steps have been shared: (a) the bipolar collector and PMOS N-well processes; (b) the NMOS N+ source/drain, bipolar emitter and collector contact steps; (c) the PMOS P+ source/drain and base contact steps. One extra processing/mask level has been introduced into the basic CMOS process flow to incorporate the bipolar P-base.

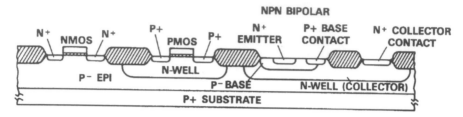

Figure 3.3 BiCMOS structure formed by the simple addition of an NPN bipolar transistor to the basic N-well CMOS process of Fig. 3.2.

From a bipolar performance standpoint, this simple approach has a number of limitations. The most significant of these is the lightly doped PMOS N-well (approximately 2K ohms/sq) that is used to form the bipolar collector. The low doping concentration leads to a large collector resistance, which limits the usefulness of the bipolar transistor. Figure 3.4 illustrates how this problem is overcome by introducing a buried N+ layer under the N-well. The buried N+ layer is first formed in the P- substrate and then capped with a 2 µm thick N-type epitaxial layer. This approach to minimizing collector resistance through the use of a heavily-doped buried N+ layer is typical of bipolar-only processes. In the case of BiCMOS processes, the use of a heavily doped N+ layer under the N-well not only reduces collector resistance but also reduces the susceptibility to latchup (section 3.2). This allows the P- epitaxial layer on the heavily doped P+ substrate to be replaced with a P- substrate for the same, or better, latchup immunity. The CMOS P-well, for the NMOS device, and the bipolar transistor are then fabricated in the thin N- epitaxial layer. The resisitivity of the N- epitaxial layer (typically 1 ohm-cm) is chosen so that it can support both the PMOS and NPN bipolar transistors. The collector series resistance is further reduced by adding a deep N+ connection to the buried N+ subcollector. Thus, with the addition of two mask levels (buried N+ and deep N+ collector), this approach merges the process steps needed to achieve low bipolar collector resistance with those required to reduce CMOS latchup susceptibility.

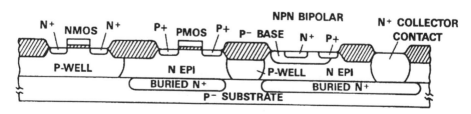

Figure 3.4 BiCMOS structure showing the introduction of a buried N+ layer and deep N+ topside contact to reduce collector resistance.

The above approach, although producing a bipolar device with much improved characteristics, still has a number of drawbacks. In particular, the packing density of the bipolar devices is limited by the P- substrate doping level that must be used to prevent punchthrough from one bipolar collector to another. Raising the doping level of the P- substrate, while allowing the bipolar devices to be more closely spaced, causes increased collector-to-substrate capacitance. Also, the N-type epitaxial layer has to be counterdoped to isolate the N-well regions and form P-wells for the NMOS devices. Counterdoping the 2 μm thick 1 ohm-cm N-type material can cause processing difficulties and a reduction in NMOS performance through mobility degradation.

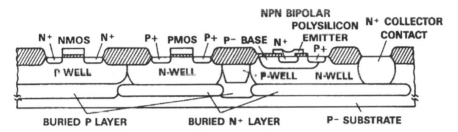

Figure 3.5 Optimized BiCMOS device structure. Key features include self-aligned P and N+ buried layers for improved packing density, separately optimized N and P-wells (twin-well CMOS) formed in an epitaxial layer with intrinsic background doping, and a polysilicon emitter for improved bipolar performance.

Figure 3.5 shows how an improvement in bipolar packing density can be achieved by self-aligning buried P layers to the buried N+ regions [3.1,3.7]. This approach allows the collector-to-collector spacing to be significantly reduced, but at the cost of increased collector-to-substrate sidewall capacitance. Figure 3.5 also shows how a twin-well CMOS structure can be implemented without heavily counterdoping the N-type epitaxial layer. Instead of depositing an N-type epitaxial layer whose doping level is determined by the bipolar and PMOS device requirements, a near-intrinsic epitaxial layer is deposited [3.8]. Self-aligned P and N-wells are then implanted into the thin intrinsic epitaxial layer, allowing independent optimization of each doping profile. For the cost of an extra mask level, the bipolar device performance can be further improved by replacing the diffused emitter (which was formed by the N+ source/drain) with a polysilicon emitter. The additional process complexity of this approach can be minimized if both the CMOS gates and the bipolar emitter utilize a common polysilicon layer. Higher bipolar performance is obtained from polysilicon emitters since shallower emitters and narrower basewidths can be achieved, reducing both transit time and parasitic capacitance (through a reduction in emitter-base junction area). Four additional mask levels (buried N+, deep N+ contact, P-base and emitter) are required to merge this higher performance BiCMOS process with a baseline CMOS flow, resulting in a total mask count of fourteen when compared with a typical ten mask double-level-metal CMOS process.

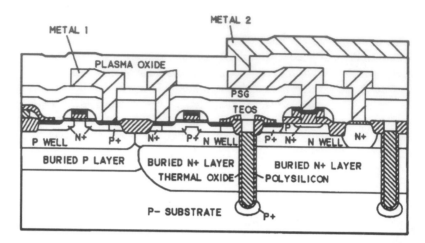

Figure 3.6 High performance submicron BiCMOS process cross-section showing trench-isolated bipolar devices and silicided gates, emitters and diffusions for low sheet resistance (~1.5 ohms/sq).

Based on the general approach shown in Fig. 3.5, a high performance submicron BiCMOS process can be implemented [3.3]. Such an approach is shown in Fig. 3.6, where the gates, emitters and diffusions have been silicided using a self-aligned TiSi$_2$ ("salicide") process, reducing sheet resistances from typically 20-50 ohms/sq to 1.5 ohms/sq. Reducing the polysilicon sheet resistance to 1.5 ohms/sq results in a bipolar device with lower emitter resistance, permitting higher emitter current densities to be realized. The silicided diffusions also reduce the NMOS and PMOS source/drain resistances, resulting in higher drive currents. Similarly, silicidation of the bipolar extrinsic base and collector diffusions provide lower series resistance and enhanced bipolar performance.

Sidewall oxide spacer technology allows the addition of lightly doped drains to the NMOS transistors to prevent the generation of hot electrons that would otherwise cause reliability problems. The P+ extrinsic base can also be self-aligned to the sidewall oxide on the edge of the polysilicon emitter, allowing smaller base area and lower collector-base capacitance. An additional enhancement is the implementation of bipolar trench isolation to increase bipolar packing density and reduce the collector-to-substrate capacitance for improved performance. Local interconnect (LI) technology can also be used to connect polysilicon gates, emitters and diffusions without the need for contact holes and metal staps. Since source/drain and base junction areas can be reduced through the use of LI, higher packing density and lower junction capacitance can be achieved, leading to increased chip packing density and performance. Finally, tungsten metallization is used to plug submicron contacts and vias to prevent metal step coverage problems and suppress electromigration. A detailed description of this type of high performance submicron BiCMOS process flow is given in section 3.7 of this chapter.

3.2 BiCMOS Isolation Considerations

BiCMOS device isolation is a key factor in determining overall circuit performance and density, and the tradeoffs between optimum speed, packing density and process complexity must be carefully considered. Since high performance bipolar NPN transistors require a heavily doped buried N+ layer, or subcollector, to lower the collector series resistance underneath the thin epitaxial collector region, the BiCMOS well formation strategy is dramatically different from standard CMOS.

In rudimentary CMOS, a single well region is diffused into an epitaxial layer which has a doping concentration appropriate for the complementary device and a thickness sufficient to avoid punchthrough between the well and underlying substrate (see Fig. 3.2). For twin-well CMOS, the N and P-well doping profiles are optimized by separate diffusions, and the background doping of the epitaxial layer is held sufficiently low ($<1 \times 10^{15}$ atoms/cm^3) to avoid the need for excessive counterdoping during N or P-well formation since this would degrade carrier mobility. In order to suppress latchup and ground bounce, a heavily-doped substrate is commonly used to shunt any voltage drops caused by substrate currents.

3.2.1 Latchup

The latchup phenomenon, generic to CMOS circuits, stems from the presence of a PNPN thyristor formed by parasitic PNP and NPN bipolar transistors which are intrinsic to the CMOS well structure. The PNP and NPN parasitic bipolar transistors which combine to form a thyristor in a twin-well CMOS inverter are highlighted in Fig. 3.7. Positive feedback can occur if one of the source/drain junctions is momentarily forward biased (for example, triggered by spurious noise, voltage overshoots, electrostatic discharge or the application of signal levels to inputs before power-up), since the collector of one transistor feeds the base of the other and vice versa. This causes a sustained high current to flow between V_{ss} and V_{DD}, resulting in the latchup condition (Chapter 4).

Apart from utilizing a heavily doped substrate for latchup suppression, alternative approaches include the use of retrograde wells to lower well resistance and effectively reduce the gain of the vertical PNP and lateral NPN device. The retrograde wells can be formed by high energy ion implantation [3.9] or by the use of buried layers (as shown in Figs. 3.4 and 3.5). With the first approach, epitaxy is not required, and a lightly doped substrate can be used to minimize the bottom wall capacitance of the complementary well region. The major disadvantages of this approach include limited sublayer doping from ion implantation damage considerations and residual defects introduced by the high energy/dose implant [3.10].

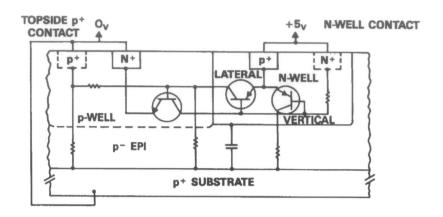

Figure 3.7 CMOS cross-section showing PNPN latchup circuit .

3.2.2 Buried Layers

The above disadvantages can be overcome by combining the retrograde concept with the epitaxial approach. For example, in Fig. 3.4, a heavily doped buried N+ region is formed in a lightly doped substrate and capped by a uniformly doped N-type epitaxial layer that serves as the PMOS N-well and bipolar subcollector. The P-well is then formed by selectively counterdoping the N-type epitaxial layer. This approach offers good control of the bipolar collector doping profile but has the disadvantage of lower electron mobility in the P-well, since significant counterdoping of the N-type epitaxial layer is required. Figure 3.5 illustrates an extension of the twin-well concept in which both the dual buried layers and overlying wells are formed using self-aligned techniques. This approach has the advantages of fewer masking levels and higher packing density, and allows independent optimization of N and P-well doping profiles, since an intrinsic background doping can be used for the epitaxial layer.

However, it should be noted that the use of buried layers, apart from adding process complexity, recesses the buried N+ region by approximately 100-200 nm, since a silicon step is required for alignment of the subsequent mask level. Additional topography is also introduced by the self-aligned twin-well process, and if this step occurs over the N-well, the focal plane of the N-well will be 200-400 nm lower than the P-well. Although present lithography tools have adequate depth of focus to achieve acceptable linewidth control over this topography, a coplanar well formation process may be required for the 0.5 μm regime. However, buried layers will still play an essential role in the BiCMOS device structure, with the thickness and doping concentration of the buried N+ region and overlying epitaxial layer normally determined by the requirements for optimum bipolar performance.

Since low collector series resistance is essential for high performance bipolar devices and latchup prevention, buried N+ subcollector doping levels exceeding $1x10^{19}$ atoms/cm^3 are typically used (corresponding to sheet resistances of 20-30 ohms/sq for a 2-3 μm deep buried layer), and the buried N+ anneal and subsequent epitaxial deposition processes must be optimized to reduce material defects and minimize autodoping. The buried N+ layer is normally formed by ion implantation and drive/anneal of arsenic or antimony, the latter having the advantage of exhibiting less autodoping during epitaxy and subsequent thermal cycles but the disadvantages of having lower solid solubility and requiring higher anneal/drive temperature for buried layer formation (1250°C as compared with ~1000°C for arsenic). During the buried N+ drive/anneal, lateral diffusion of the dopant effectively limits the spacing between adjacent subcollector regions and determines the ultimate N-well to N-well spacing. A boron punchthrough stopper is normally implanted between the subcollector regions to provide more punchthrough margin, and this can either be patterned or done in a self-aligned manner [3.1]. Alternatively, a more heavily doped substrate can be utilized, but this has the disadvantage of increasing the N-well to substrate bottomwall capacitance.

3.2.3 Epitaxy and Autodoping

The twin buried layers inherent to the BiCMOS device structure complicate the epitaxy process, since vertical and lateral autodoping effects must be considered for both N and P-well regions. For given buried layer dopant concentrations and species, epitaxial growth conditions such as the reactant, temperature, pressure, etch/bake cycles and growth rate must be specifically tailored to minimize autodoping in order to achieve shallow well profiles (1-1.5 μm typical) and avoid excessive counterdoping of adjacent well regions with opposite polarity [3.11]. For lighter background doping levels (such as intrinsic-doped epitaxy), autodoping becomes a more serious concern, especially for thinner epitaxial layers.

For arsenic-doped buried layers, reduced pressure (50-100 Torr) has been shown to dramatically reduce autodoping and also decrease the pattern shift associated with the epitaxial growth process [3.12]. However, boron autodoping becomes more pronounced at lower pressures [3.13], so the epitaxial process conditions must be compromised to suppress both boron and arsenic autodoping. Boron redistribution can also be reduced by using a lower-temperature epitaxy process (< 1100°C, as opposed to typical epitaxial growth at 1100-1200°C), but again, a compromise is required since arsenic and antimony autodoping effects increase at lower temperatures (until dopant reabsorption starts to dominate over evaporation into the gas phase) [3.11, 3.13]. Thus, temperature and pressure must be judiciously chosen to minimize autodoping from both the N and P type buried layers.

Lateral autodoping can also be reduced if the time and temperature of the prebake period (the stabilization period between in situ etching of the surface and actual deposition of the epitaxial layer) is increased, since the buried layer dopants diffuse to the surface and evaporate into the gas stream during this cycle [3.14]. However, the pre-deposition cleaning sequence used to preclude defects (hydrogen reduction to remove residual oxides, HCl silicon etch and hydrogen prebake) must also be optimized to avoid overly depleting the buried layer dopants, since electrical isolation, latchup and bipolar performance would suffer.

3.2.4 Parasitic Junction Capacitance

When designing a BiCMOS process flow, it is important to consider the advantages and disadvantages of the various approaches to buried layer and well formation in order to match device isolation and performance requirements with process capability. BiCMOS well design is, to a large extent, dictated by the DC characteristics of the CMOS and bipolar device, since epitaxial doping and thickness must be tailored to yield specified breakdown voltages (BV_{CBO} for the bipolar transistor and BV_{DSS} for the MOSFETs). Performance tradeoffs can be attributed primarily to the merging of these dictated well dopings with the various parasitic capacitances of the device.

For the BiCMOS well structure shown in Fig. 3.5, the major parasitic capacitance-to-substrate is encountered at the sidewall. Obviously, this capacitance will increase as higher punchthrough doses are used to isolate tighter N-well to N-well spaces. By calculating the avalanche and punchthrough voltages as a function of P-layer concentration for a given N-well separation, the range of boron concentrations required for isolation can be estimated. In Fig. 3.8, avalanche and punchthrough-limited subcollector-to-subcollector breakdown voltage is plotted as a function of buried P-layer dopant concentration for two different as-implanted buried N+ spacings (a lateral N+ diffusion of approximately 2 μm is assumed during the buried N+ anneal). Note that as expected, tighter subcollector spacings require higher P-layer doping levels to maintain a given breakdown voltage, and correspondingly higher values of sidewall capacitance (see right-hand scale) are incurred.

3.2.5 Trench Isolation

Both collector to P-well sidewall capacitance and subcollector spacing can be minimized by using the trench isolated structure illustrated in Figure 3.6. Although several techniques [3.15-3.17] have been proposed for achieving trench isolated structures, including the novel use of selective epitaxy [3.18], the most commonly used method employs an oxide-isolated polysilicon-filled trench (shown in Fig. 3.6), which penetrates beyond the n+ buried layer to minimize sidewall capacitance and prevent latchup. Typical sidewall capacitance for the trench structure is only on the order of 0.1fF/μm^2, albeit at the cost of additional process complexity. For a BiCMOS scenario, sidewall passivation also becomes an issue, since leakage currents have been observed along the P-well/trench interface [3.19]. Still, if the sidewall leakage can be reduced, trench isolation becomes a viable approach for the aggressive scaling of BiCMOS well regions, especially since trench-isolated bipolar structures have already been demonstrated in production [3.20].

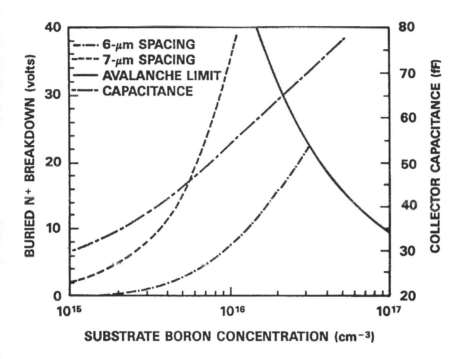

Figure 3.8 Avalanche and punchthrough-limited subcollector-to-subcollector breakdown voltage plotted as a function of buried P-layer dopant concentration for two different as-implanted buried N+ spacings. Model assumes an NPN transistor with perimeter = 33 μm and sidewall area = 100 μm^2, with a lateral diffusion of ~ 2 μm during the buried N+ anneal.

3.2.6 Active Device Isolation

A similarly aggressive approach is also required for the isolation of active device regions, since standard LOCOS technology exhibits too much active area encroachment, or "bird's beak", with respect to typical design rules used for submicrometer technologies (active region on the order of 2-2.5 μm). For BiCMOS structures, it is also important to minimize the thermal cycle once the wells are in place, since up-diffusion of the buried layers will lead to increased diode capacitance (MOS source/drains and bipolar collector-base), higher MOS transistor body effect, and lower bipolar collector-base breakdown voltage. At least two approaches can be taken to limit the diffusion of the buried layers during subsequent processing.

First, high energy implantation can be used to introduce the well doping after thick field oxidation, with the range of the well implants corresponding in proximity to the field oxide/silicon interface. After annealing the implant, a retrograde well structure is formed, which also provides for enhanced channel stop doping near the field oxide/silicon interface. Using this technique, 1μm active region widths and 2 μm N+ to P+ isolation have been demonstrated for CMOS devices [3.9]. Disadvantages include the need for high energy ion implantation (~1 MeV), concern over ion implantation-induced damage and the difficulty in implementing the process in a self-aligned manner, since very thick implantation masks are required. Higher diode capacitance and MOS transistor body effect may also be incurred if the peak N-well and P-well concentrations lie near the respective P+ and N+ source/drain junctions.

A more conventional approach to limiting the thermal cycle associated with field oxidation involves the use of high pressure oxidation. High pressure field oxidation not only minimizes the diffusion of the buried layers and wells, but also reduces the segregation of boron at the field oxide/silicon interface. The latter attribute permits a lower channel stop dose to be used in preserving adequate NMOS thick field oxide threshold voltage (>10 V). Thus, for an unmasked channel stop implant, there is less counterdoping of the N-well region by the boron channel stop, and a correspondingly higher PMOS thick field oxide threshold voltage can be achieved. Using high pressure oxidation and a modified LOCOS technique known as poly buffer LOCOS (PBL), BiCMOS circuits with 1.2 μm active device region isolation and 3 μm N+ to P+ spaces have been demonstrated [3.3].

The PBL technique is one of several recent approaches [3.21-3.23] to limiting LOCOS active area encroachment. In this particular approach, a polysilicon buffer layer is inserted between the pad oxide and the overlying nitride oxidation mask of traditional LOCOS technology. The polysilicon serves as an additional stress relief layer, allowing a thinner oxide and thicker nitride to be used during field oxidation, which reduces encroachment without inducing defects. Key steps in the PBL process are illustrated in Fig. 3.9. Section 3.7 will address the implementation of PBL, and other key aspects of BiCMOS isolation described above, within the framework of an integrated BiCMOS process flow.

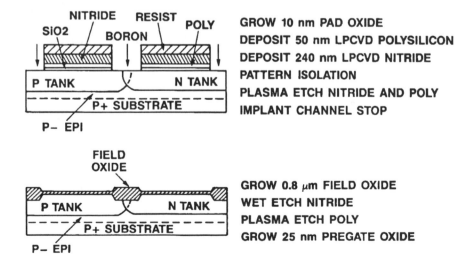

Figure 3.9 Key steps in the polysilicon buffer LOCOS (PBL) process.

3.3 CMOS Well and Bipolar Collector Process Tradeoffs

The optimization of submicrometer BiCMOS well profiles is a multifaceted problem due to the conflicting requirements of the mixed bipolar and CMOS technologies. Since the N-well (or N-epitaxial layer for P-well CMOS) provides the foundation for the bipolar NPN and PMOS transistor, most of the critical process tradeoffs involve the N-well profile. However, key issues such as isolation, diode capacitance and mobility degradation in counterdoped P-wells must also be considered, especially in CMOS-intensive circuits where the NMOS transistor performance plays a dominant role.

From a bipolar standpoint, the N-well profile is a compromise between a thinner, more heavily doped collector region (which minimizes base pushout, or Kirk effect, and leads to a higher f_T and current handling capability) and a thicker, more lightly doped collector (which provides a higher collector-base breakdown voltage). To this extent, optimization of the N-well collector doping profile is a function of specific design requirements, with the minimum current drive capability, f_T and BV_{CBO} influencing the choice of epitaxial layer thickness and N-well doping concentration. These tradeoffs are discussed in more detail in section 3.5.

From a CMOS perspective, the well profiles must be optimized with respect to key MOSFET parametrics which include (1) threshold voltage, (2) punchthrough voltage, (3) source/drain diode capacitance and (4) body effect. P and N-well surface concentrations largely determine the threshold voltages of the NMOS and PMOS transistors, respectively, and surface doping levels are typically adjusted to achieve a given V_T in conjunction with a blanket boron threshold voltage implant. Below the surface, the well concentration must be sufficient to provide adequate protection against punchthrough, a consideration that drives well concentrations higher as gate lengths shrink. However, MOS performance is a strong function of source/drain diode capacitance and body effect, and these parasitics are minimized by lowering the well concentration.

Thus, a compromise is again required that must, for the BiCMOS case, also include the bipolar device requirements discussed above. In this respect, it should be noted that the CMOS and bipolar devices do share some common goals, such as the need to minimize diode capacitance (N+/P+ source/drains for CMOS and collector-base for bipolar) and the need to maintain sufficient N-well concentration at the field oxide-silicon interface to provide adequate thick field oxide threshold voltage for PMOS and bipolar transistor isolation. However, while general boundary conditions have been qualitatively discussed [3.24], actual well profiles tend to depend strongly upon the choice of technology, i.e., N-well, P-well or twin-well, in addition to the design requirements.

As one example of BiCMOS well design, consider the case of an 0.8 μm BiCMOS technology [3.3] which utilizes N+/P+ buried layers and twin-wells diffused into a thin (1.0-1.5 μm) intrinsically-doped epitaxial region. Figure 3.5 shows a representative device cross-section for this process, which is described in more detail in section 3.7. For purposes of comparison, the BiCMOS well design is contrasted with a comparable 0.8 μm CMOS-only process [3.5] in order to highlight the key differences in well profiles. The CMOS flow also utilizes a twin-well process, but the wells are diffused into a thick (5-6 μm) P- epitaxial layer grown on a P+ substrate. The well diffusion is relatively long for the CMOS case (500 min at 1100°C), and produces deep N and P wells with relatively constant doping profiles. This is illustrated in Figs. 3.10 and 3.11, where the respective source/drain doping profiles have been included.

Measured P+/N-well and N+/P-well doping profiles for the BiCMOS process are plotted in Figs. 3.12 and 3.13 respectively. Note that both the N and P-well BiCMOS profiles exhibit a steeper gradient than their CMOS-only counterparts, and that the zero-bias depletion widths are correspondingly wider for the BiCMOS wells, leading to lower capacitance. Due to bipolar device requirements (minimum base pushout, etc.), the BiCMOS N-well purposely has less gradient than the P-well, and this tailoring is accomplished by a combination of low and high energy phosphorus implants. One negative consequence of a steeper bipolar N-well profile is that it leads to higher collector series resistance; however, a deep N+ diffusion, (Fig. 3.4), is typically used to provide a low resistance path to the buried N+ subcollector, thereby eliminating the problem.

The impact of lower diode capacitance on circuit speed is best illustrated by comparing the performance of CMOS inverter chains fabricated with the 0.8 μm BiCMOS and CMOS-only processes. In Fig. 3.14, CMOS inverter delay per stage is plotted as a function of supply voltage for circuits fabricated in BiCMOS and CMOS-only processes. The drive currents were matched in each case at 0.44 mA for the W/L = (1.2 μm/0.8 μm) NMOS transistors and 0.21 mA for the W/L = (1.4 μm/0.8 μm) PMOS transistors. Note that for identical test conditions at 5V , the delay is approximately 20% less for CMOS inverter chains fabricated with the BiCMOS process. This increase in speed is consistent with the significant reduction in bottomwall capacitance measured for the BiCMOS case.

The dashed curve in Fig. 3.14 shows the measured delay per stage for a CMOS inverter chain fabricated in the BiCMOS process but designed with a 25% reduction in the drain areas. This area reduction was achieved by the use of a local interconnect technology, which is discussed in more detail in Section 3.6. By utilizing local interconnect technology, an additional 11% decrease in inverter chain delay is obtained, further demonstrating the importance of minimizing diode capacitance to optimize circuit performance.

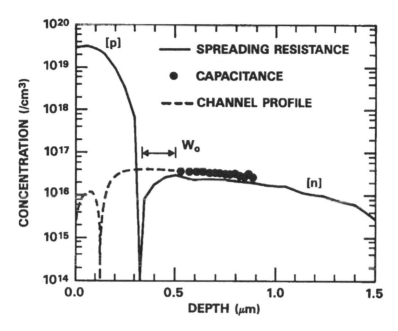

Figure 3.10 Measured P diode in N-well characteristics for the CMOS-only process. The triangles show the profile obtained from P diode capacitance measurements. W_0 is the zero bias depletion depth and the channel profile under the gate is simulated. A and B define the edge of the channel depletion layer at 0 V and + 2.5 V N-well bias.

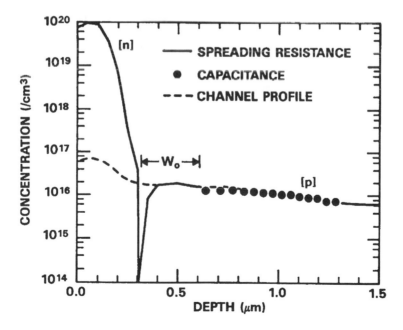

Figure 3.11 Measured N diode in P-well characteristics for the CMOS-only process (nomenclature as in Fig. 3.10).

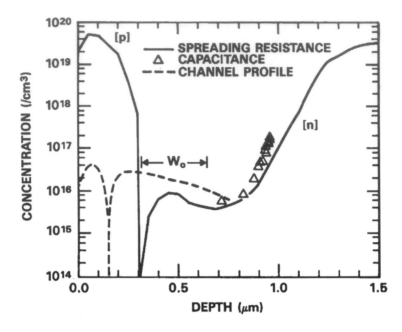

Figure 3.12 Measured P diode in N-well characteristics for the BiCMOS process (nomenclature as in Fig. 3.10).

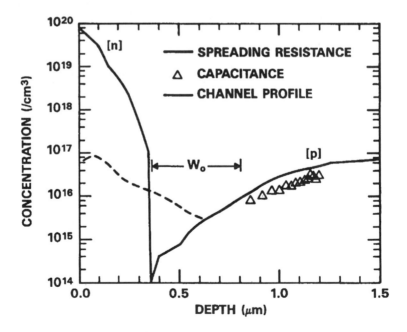

Figure 3.13 Measured N diode in P-well characteristics for the BiCMOS process (nomenclature as in Fig. 3.10).

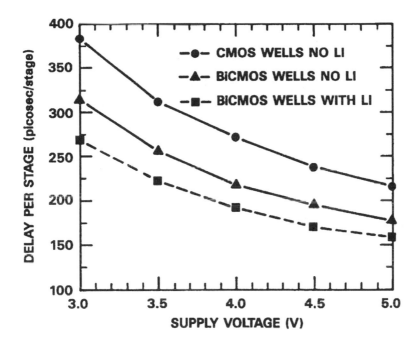

Figure 3.14 A comparison of measured performance for CMOS inverter chains fabricated in CMOS-only and BiCMOS processes. For all cases, NMOS W/L=1.2 μm/0.8 μm and PMOS W/L=1.4 μm/0.8 μm. Measured inverter chain performance for CMOS devices utilizing a local interconnect in the BiCMOS process is also shown (dashed line).

3.4 CMOS Processing Considerations

Two of the most important considerations in CMOS transistor design for a submicron BiCMOS process are high performance and reliability. Since VLSI BiCMOS circuits tend to be CMOS-intensive due to power dissipation limitations, the CMOS transistors must be designed for high speed as it is not always possible to use bipolar circuits in critical speed paths. NMOS transistor reliability becomes a major issue in the submicron regime because of hot carrier generation, which can lead to threshold voltage and drive current shifts [3.25].

In a high performance submicron BiCMOS process, the CMOS devices are typically fabricated using a twin-well CMOS approach [3.5]. The P-well doping and concentration gradients are optimized for the NMOS transistor and the N-well profile is chosen to simultaneously support the requirements of both the PMOS and bipolar NPN devices. Typically, the NMOS and PMOS gates are polysilicon and may be doped at the same time as the bipolar polysilicon emitter. The doping is normally provided by a high dose phosphorus or arsenic ion implantation in the range of 5×10^{15} to 1×10^{16} atoms/cm^2. The N-type polysilicon gates result in a surface channel NMOS device and a buried channel PMOS transistor.

In the following sections the various process design issues that impact CMOS transistor performance and reliability are reviewed. These include threshold voltage adjustment, drain design and channel doping profile. Finally, the relationship between manufacturing control and device performance is discussed. Source, drain and gate silicidation is covered in the section on interconnections.

3.4.1 CMOS Threshold Voltage Considerations and Adjustment

To ensure adequately low source-to-drain leakage currents at high temperature (125 °C), the threshold voltage, defined as the gate voltage at zero drain current (from a numerical regression of gate voltage and drain current above the threshold voltage), is typically 0.6 V for 0.8 μm gate length transistors with a drain voltage of 5 V. The NMOS and PMOS threshold voltages can be adjusted to this value by implanting boron into both the NMOS and PMOS channel regions. The boron causes the NMOS threshold voltage to increase above the value determined by the P-well doping, since it increases the P-type concentration in the channel region. The PMOS threshold voltage, on the other hand, becomes less negative than the value determined by the N-well, since the boron reduces the N-type concentration in the channel region. Typically, the boron counterdopes the N-well just below the surface and forms a thin buried P-channel. The PMOS buried channel is shallow enough to be pinched-off with zero volts on the gate, and has adequate subthreshold turnoff characteristics.

Through a judicious choice of N-well and P-well surface concentrations and boron dose, the NMOS and PMOS threshold voltages can be simultaneously adjusted using a single unmasked boron ion implantation. The boron required to set the threshold voltages (approximately 1.5×10^{12} atoms/cm^2), is normally implanted through a pregate gate oxide, which is removed prior to gate oxidation. Typical gate oxide thicknesses range from 17nm to 20 nm for an 0.8 μm process. The polysilicon gate electrode (approximately 0.5 μm thick) is deposited immediately following gate oxidation, protecting the polysilicon-oxide interface. This procedure preserves the gate oxide integrity, since the gate oxide is not exposed to any photoresist patterning or ion implantation steps.

3.4.2 Source/Drain Processing and Channel Profiles

As the NMOS gate length is scaled to the submicron regime, hot electron generation due to impact ionization becomes a critical reliability hazard. To overcome this problem and to allow circuits to operate at a 5 V supply voltage, special attention must be given to how the drain junction is formed. The generation of hot electrons, that can subsequently be injected into the gate oxide and cause shifts in threshold voltage and transconductance, can be greatly decreased by reducing the electric field at the drain junction. The reduction in the electric field is usually accomplished by introducing a lightly doped drain extension (LDD) or grading the concentration profile of the drain diffusion [3.26-3.28]. In both approaches, the drain depletion region extends further into the drain diffusion, thereby reducing the electric field and also increasing the drain-to-source punchthrough voltage.

Figure 3.15 shows NMOS LDD and double-diffused graded drain structures. Both techniques rely on the use of sidewall oxide spacer technology. The sidewall oxide spacer is formed on the vertical edges of the polysilicon gate electrode by depositing a high quality oxide, typically 300 nm of TEOS, and then removing it with an anisotropic etch. The anisotropic nature of the etch results in oxide "filaments" being left along the edges of the vertical polysilicon gates, with a bottom width approximately equal to the deposited thickness. To ensure a high quality interface under the sidewall oxide, it is important that the thermally grown gate oxide remain in place after the polysilicon gate etch. This requires a polysilicon etch with extremely high selectivity to oxide, since the gate oxide is typically 20 nm or less.

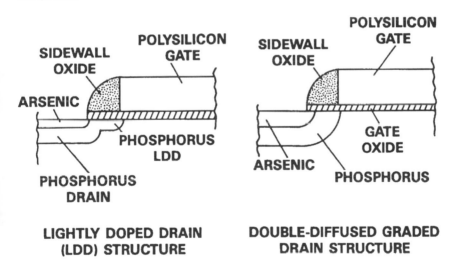

Figure 3.15 NMOS LDD and double diffused graded drain structures.

In the case of the LDD approach to reducing the generation of hot electrons, a light N-type ion implantation, normally $1-2 \times 10^{13}$ atoms/cm^2 of phosphorus, is self-aligned to the polysilicon gate edge prior to sidewall oxide deposition. After formation of the sidewall oxide filament, the more heavily doped source/drain ion implantation is performed, self-aligned to the edge of the sidewall oxide. This ion implantation is typically arsenic or a mixture of arsenic and phosphorus at a dose of about 5×10^{15} atoms/cm^2. Hence, the sidewall oxide acts as a mask against the high dose source/drain ion implantation, preventing it from entering the more lightly doped region. A cross-section of an 0.8 μm NMOS transistor fabricated using the LDD process is shown in Fig. 3.16, while the doping profiles along the

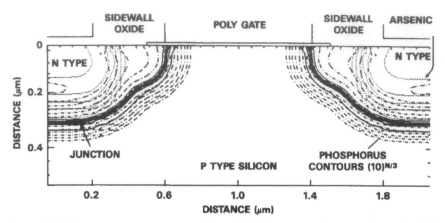

Figure 3.16 Process cross-section of an 0.8μm NMOS transistor fabricated using the LDD process with a phosphorus plus arsenic source/drain.

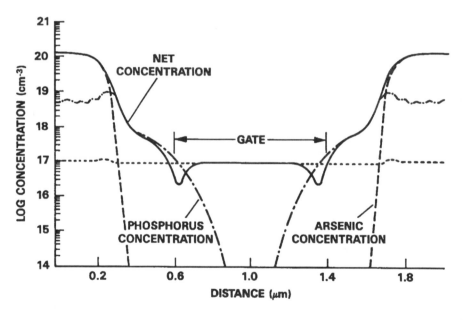

Figure 3.17 Doping profiles along the channel of the NMOS structure of Fig. 3.16.

channel are shown in Fig. 3.17. Respective junction depths for the LDD and heavily doped source/drain regions are typically 0.18 μm and 0.3 μm. Note that although the focus has been on the drain structure, the source also has an "LDD" extension due to the symmetry of the MOSFET. While this effect is useful in circuits where the source and drain of a device will be reversed, for example a pass gate, it normally degrades the performance of the device through added source resistance, which results in reduced transconductance. The increased resistance at the drain terminal raises the value of the drain saturation voltage.

For the case of the graded drain structure, it is common practice to form the sidewall oxide filament along the edge of the gate before introducing any source/drain dopants. After sidewall oxide formation, arsenic and phosphorus are implanted into the source/drain regions with typical doses of 5×10^{15} and 1×10^{15} atoms/cm^2, respectively. The phosphorus is then driven ahead of the arsenic underneath the sidewall oxide to "grade" the drain doping profile and thereby reduce the electric field. The advantage of this technique is that it can be introduced into a CMOS process flow without any extra masking steps, whereas the LDD approach typically uses the N+ source/drain mask to also pattern the LDD, increasing the total number of masking steps by one. However, the LDD approach is more scalable since it offers better electrical channel length control and device isolation. Improved electrical channel length control follows since the only etching step that determines the channel length is the polysilicon gate etch. In the graded drain approach, the variation in sidewall oxide thickness (which is a function of the deposited oxide thickness, oxide etching and polysilicon gate edge profile) also contributes to determining the electrical gate length.

The LDD approach offers better isolation scalability because the source and drain junction depths can be made shallower, since they do not have to be driven under the sidewall oxide to make contact with the active channel region. Shallower source/drain junctions permit closer separation of active regions and tighter isolation pitch. A further scaling advantage of the LDD structure is that it results in a source/drain junction depth immediately adjacent to the active channel that is approximately the same depth as the highly doped channel region. The high channel doping results from the shallow boron ion implantation that is used to adjust the threshold voltage. Because the LDD's are only 0.18 μm deep, the boron threshold voltage implant also acts as an N+ source/drain punchthrough stopper, thereby improving short channel characteristics and obviating the need for a deep punchthrough implant. The NMOS device body effect is also kept low since the channel doping profile drops off rapidly with depth to the lower P-well background concentration (approximately 1×10^{16} atoms/cm^3). This design also results in low junction bottomwall capacitance, since the heavily doped source/drain regions are much deeper than the peak channel doping.

When a buried channel PMOS device is used, it is unnecessary to use either a PMOS LDD or to drive the P+ source/drain totally underneath the sidewall oxide to provide a link with the active channel region. This follows because the buried channel, formed by the boron ion implantation used to adjust the PMOS threshold voltage, extends beyond the gate edge underneath the sidewall oxide, creating a buried channel LDD, or BCLDD [3.29]. The structure of an 0.8 μm buried channel PMOS device is depicted in Fig. 3.18, and the doping profile along the channel in Fig. 3.19. Although the existence of the BCLDD makes it unnecessary to drive the P+ source/drains totally underneath the sidewall oxide, it is preferable to drive the source and drains to within approximately 50 nm of the active channel. This follows because of the rapid increase in source/drain resistance that is associated with longer BCLDD's, caused by the relatively light dose used for threshold voltage adjustment.

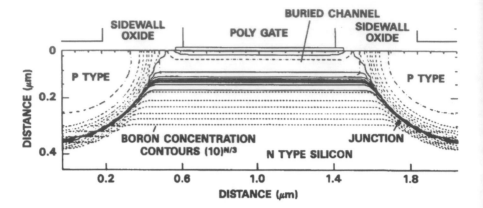

Figure 3.18 Process cross-section of an 0.8μm PMOS transistor showing the buried channel region that serves as a source/drain extension under the sidewall oxide.

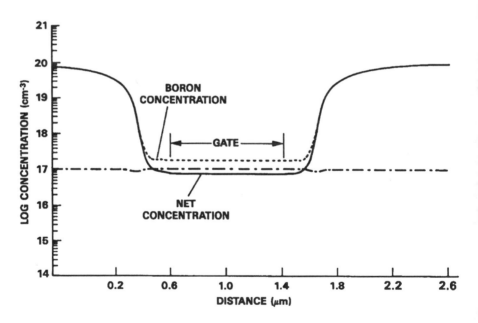

Figure 3.19 Doping profile along the channel of the PMOS structure down in Fig.3.18.

Although unnecessary for PMOS buried channel devices of the 1 μm class, boron PMOS "LDD's" become important as dimensions shrink toward 0.5 μm, where it becomes necessary to change the device from buried to surface channel and minimize gate-induced diode leakage. A change in the PMOS channel design is needed due to difficulties encountered in controlling the subthreshold turnoff and leakage characteristics of ultra-short buried channel PMOS transistors. In the

PMOS case, the "LDD" is used primarily to make the source/drain junctions shallower immediately adjacent to the active channel region and to provide electrical conductivity between the active channel region and the P+ source/drain. This is in contrast to the NMOS device, where the LDD is used primarily to reduce hot electron generation. The PMOS "LDD" is formed in a similar manner to the phosphorus NMOS LDD structure. However, it is unnecessary to use another mask to prevent the light dose of boron, typically 8×10^{12} atoms/cm^2, from entering the NMOS LDD and source/drain regions, since the masked higher dose NMOS phosphorus LDD ion implantation can be used to counterdope the boron.

3.4.3 Relationship Between Transistor Performance and Manufacturing Control

Another important process design consideration for submicron CMOS is reliable performance over the range of manufacturing tolerances. For example, a typical patterned 0.8 μm NMOS transistor may have an electrical gate length of 0.7 μm and a 3 sigma manufacturing tolerance on patterned gate length of +/- 0.12 μm. Consequently, the shortest electrical channel length the process must support is 0.58 μm. At this gate length, the process must result in CMOS devices that have adequate hot carrier reliability, threshold voltage control, subthreshold turnoff/ leakage and breakdown voltages. For the process to support the minimum gate length device parameters, the doping levels must be raised above that necessary

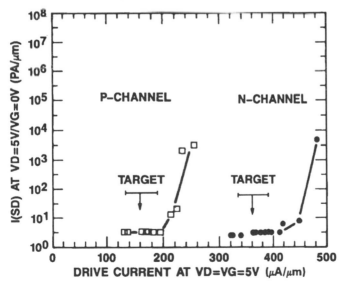

Figure 3.20 NMOS and PMOS drive current, defined as the drain current for 5V applied to the gate and drain, versus device leakage in the off state with 0 V on the gate and 5 V on the drain. The parameter is gate length, with a target value of 0.8 μm.

for the nominal device length. Consequently, the performance of the nominal gate length device tends to be degraded through increased source/drain junction capacitance and body effect caused by the higher channel doping required to prevent punchthrough and subthreshold leakage. There is, therefore, a tradeoff between manufacturing control and nominal device performance. The poorer the manufacturing control, the shorter the gate length that must be supported by the process and therefore the higher the doping levels that must be used. The importance of manufacturing control on patterned gate length is highlighted in Fig. 3.20. This figure shows a plot of NMOS and PMOS drive current, defined as the drain current for 5 V applied to the gate and drain, versus device leakage in the off state with 0 V on the gate and 5 V on the drain. The parameter is gate length. The device design has been optimized for a patterned gate length of 0.8 μm with a manufacturing 3 sigma control on patterned gate length of +/- 0.12 μm. As the -3 sigma manufacturing control limit is reached, the device leakage starts to rapidly increase. High levels of device leakage would likely cause circuits to be nonfunctional or fail a particular circuit parameter, such as standby power.

3.5 Bipolar Process Options and Tradeoffs

As discussed in section 3.1 of this chapter, the primary approach to realizing high performance BiCMOS technologies has been the addition of bipolar processing steps to a baseline CMOS process. The major challenge associated with this approach is how to simultaneously obtain high performance bipolar and CMOS transistors for a minimal increase in process complexity compared to the conventional CMOS technology. Cross-sections of various bipolar structures that have been integrated into CMOS processes are shown in Figs. 3.21-3.24. In the following subsections, the options and tradeoffs involved in integrating these bipolar devices into a CMOS flow are reviewed.

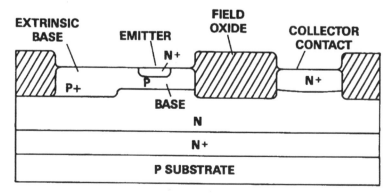

Figure 3.21 Cross-section of a bipolar transistor with a directly implanted emitter. In a BiCMOS process, the N+ collector contact and P+ extrinsic base are formed by the respective N+/P+ source/drain implants.

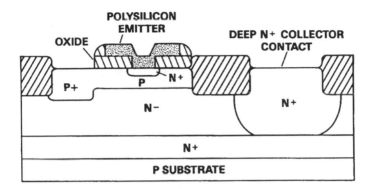

Figure 3.22 Cross-section of a bipolar transistor having a single level polysilicon emitter with an underlying protective oxide. A deep N+ collector implantation/diffusion is also used to reduce the collector series resistance.

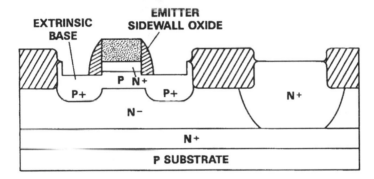

Figure 3.23 Cross-section of a bipolar transistor with a self-aligned single level polysilicon emitter. Extrinsic base diffusions are offset by a sidewall oxide to maintain adequate emitter-base breakdown voltage.

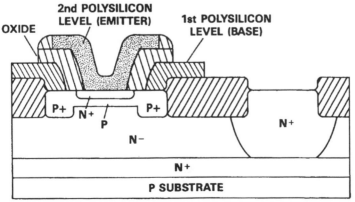

Figure 3.24 Cross-section of a bipolar transistor utilizing two levels of polysilicon. P+ polysilicon is used to contact the base region and N+ polysilicon serves as the source for the self-aligned polysilicon emitter.

3.5.1 Base Design Options

The usual measure of the bipolar transistor's AC performance is the unity gain cutoff frequency f_T, and the base width is a dominant factor in determining the magnitude of f_T. Therefore, it is normal practice to ion implant the intrinsic base region and minimize subsequent thermal cycles in order to achieve a narrow base width and maximize NPN performance. The base profile must also be optimized to avoid collector-emitter punchthrough or low emitter-base breakdown voltage. For BiCMOS processes which utilize bipolar structures similar to those shown in Figs. 3.21-3.23, the extrinsic base region is typically formed by the P+ source/drain implant.

At the same time the width of the base is being decreased to realize higher f_T, the base resistance is increasing, thereby degrading performance. By using self-alignment schemes it is possible to minimize the spacing between the extrinsic base and the emitter while also achieving a four-sided base structure [3.30], thereby reducing the base resistance. In this structure, shown in Fig. 3.24, a first layer of P+ polysilicon forms the extrinsic base electrode. After opening a window in the P+ polysilicon and forming a sidewall oxide spacer, a second polysilicon layer is deposited to form the emitter. In addition to minimizing base resistance, self-aligned transistors tend to have smaller parasitic areas, such as the extrinsic base, resulting in lower parasitic capacitance for the transistor. Another approach to minimizing the extrinsic base capacitance is to use a local interconnect for forming contacts to the base region (Section 3.6). Thus, the base junction area can be reduced since it no longer has to accommodate metal contacts.

3.5.2 Emitter Design Options

There are several methods of forming the emitter of the bipolar transistor, and the choice of emitter design usually involves a tradeoff between transistor performance and process complexity. The choices range from a directly ion implanted emitter (Fig. 3.21) to a single or double level polysilicon process, where the emitter is formed by dopant out-diffusion from the polysilicon [3.3,3.31-3.34]. While the ion implanted emitter approach is not scalable to high performance applications, it is satisfactory for applications where the f_T requirement is less than 5 GHz. For example, such an approach has been used to realize a 32 ns access time for a 1 Mbit BiCMOS DRAM [3.31,3.35].

Polysilicon emitters are used in order to obtain higher performance from the bipolar transistor. In a single level polysilicon process, the emitter polysilicon is the same layer that is used to form the MOS transistor gates (Fig. 3.22). Either phosphorus or arsenic is used to dope the emitter polysilicon, with arsenic being the choice when an f_T greater than 10 GHz is required. One of the drawbacks of using polysilicon emitters is the control of the interface between the polysilicon and single crystal silicon [3.36-3.38], since interfacial oxides can result in high emitter resistances. Phosphorus is less sensitive to interfacial layers than arsenic and yields lower emitter resistances. On the other hand, arsenic is more suited to forming the shallow emitter diffusions that are required for high performance NPN transistors.

In a double polysilicon process (as shown in Fig. 3.24), a first layer of P+ polysilicon is used to form the base electrode that can also be used to form low sheet resistance resistors. A second polysilicon layer forms the emitter of the transistor, self-aligned to the extrinsic base. In an alternate scheme of forming a narrow base for the NPN transistor using a double polysilicon process, the base is formed by diffusion of boron from the emitter polysilicon followed by diffusion of arsenic from the polysilicon to form the emitter [3.34,3.39]. NPN transistors with an f_T of 16 GHz have been fabricated using this approach in a BiCMOS process.

The emitter anneal is typically combined with the CMOS source/drain anneal for process simplicity. The base, emitter, and collector contacts are silicided at the same time as the source/drain regions and gates of the CMOS transistors in order to reduce the resistance of these regions.

3.5.3 Collector Design Options

Under high capacitance loads, the BiCMOS gate delay is strongly dependent on the collector resistance of the bipolar transistor [3.40]. Therefore, a key aspect to obtaining high performance BiCMOS circuits is to minimize the bipolar collector resistance. Additionally, high collector resistance can result in the transistor going into saturation at high collector currents, which can induce latchup by injecting holes into the substrate. The simplest way to form the collector contact is by using the N+ source/drain implant, as shown in Fig. 3.21. However, the lightly doped N-well between the bottom of the source/drain region and the buried N+ region can result in a very high collector resistance. By using a separate masking level and ion implantation for the collector contact, a lower resistance path to the buried layer can be obtained, as shown in Figs. 3.22-3.24. The disadvantage of this approach is that a relatively large collector-to-base spacing must be maintained to prevent degradation of the collector-base breakdown voltage, since significant lateral diffusion of the deep N+ will occur.

For the submicron regime, the large lateral diffusion of the deep N+ collector contact becomes a limiting factor in the scalability of the bipolar transistor. In order to obtain a scalable, low resistance path to the buried N+, a deep N+ polysilicon plug contact (Fig. 3.25) can be used [3.41]. By using in situ doped polysilicon, a doping level of greater than 1×10^{19} atoms/cm^3 can be maintained in the buried N+ plug, while at the same time a sidewall dielectric eliminates lateral diffusion and permits scaling of the transistor structure.

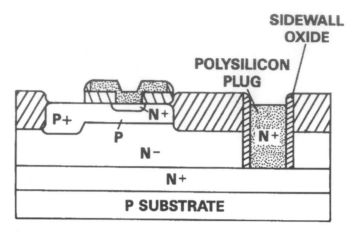

Figure 3.25 Cross-section of a bipolar transistor with a N+ polysilicon plug used to minimize the collector series resistance.

A further parameter that needs to be optimized for minimum collector resistance and maximum transistor performance is the epitaxial layer thickness. Use of a thinner epitaxial layer can increase both the f_T and the maximum current handling capability of the device before base pushout, at the expense of decreasing BV_{CEO} [3.1]. Although decreasing the epitaxial layer thickness can improve the bipolar performance, care must be taken since a thinner epitaxial layer can result in higher diode capacitance for the CMOS transistors, as previously discussed in section 3.3.

An alternative method of increasing the collector current density before base pushout occurs is to increase the N-well concentration [3.42]. However, the overall circuit performance may be degraded due to higher collector-base capacitance for the bipolar transistor, and higher diode capacitance for the CMOS transistors. This can be largely overcome by adding a phosphorus implant through the emitter window to selectively increase the well concentration in the emitter region only [3.43]. This results in a minimal increase in collector-base capacitance while serving to prevent base pushout. A further advantage is that the additional well doping cuts off the ion implantation tail of the base, thereby decreasing the base width to achieve a higher f_T for the transistor. Since the higher collector doping is localized under the bipolar emitter, this scheme can be applied to a BiCMOS process flow such that there is no impact on the performance of the CMOS transistors. Finally, as BiCMOS technology matures and enters the submicron regime, many CMOS and bipolar process steps are being merged in a way that enables high performance bipolar transistors to be realized in a baseline CMOS process, without an unreasonable increase in process complexity.

3.6 Interconnect Processes for Submicron BiCMOS

3.6.1 Silicidation

As BiCMOS is scaled to 1 µm gate lengths and below, it becomes increasingly important to minimize the series source, drain, emitter, base and collector resistances in order to realize the full performance advantages of scaled devices [3.45]. For MOS transistors, the channel conductance increases as the gate length decreases, and unless the source series resistance is reduced, the saturation transconductance will be degraded. For an NMOS transistor with an effective channel length of about 0.5 µm, the total source/drain series resistance must be reduced to approximately 1.2K ohm-µm, or the saturated transconductance will be degraded by more than 10%. Figure 3.26 highlights the impact of source and drain series resistance on device gain as the gate length is scaled. Unfortunately, the source/drain resistance tends to increase as gate lengths are scaled, since it also becomes necessary to scale the junction depths to avoid short channel effects and isolation leakage. Typically, shallow junction depths have been obtained by reducing the doping concentration of the diffusions which, particularly for the PMOS transistor that uses boron as the dopant, result in higher source and drain resistance. In the case of the NMOS transistor, the situation is less critical since arsenic can be used to create shallow, highly doped diffusions. Consequently, as Fig. 3.26 shows, the degradation in performance is most serious for the PMOS transistor. For the case of scaling bipolar transistors, decreasing the base, collector and emitter resistances all become critical issues in realizing higher operating frequencies and improved collector current density.

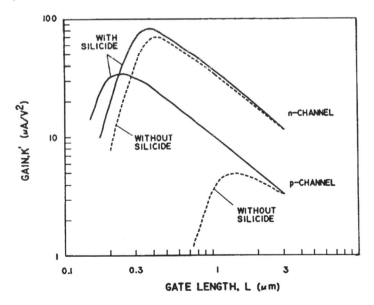

Figure 3.26 Impact of source and drain series resistance on the performance of scaled PMOS and NMOS devices.

To meet the submicron BiCMOS requirements of lower source/drain, gate, emitter, base and collector series resistances, a technique known as self-aligned silicidation, or salicide, has been developed [3.46]. This technology results in the simultaneous reduction of polysilicon and diffusion sheet resistances from typically 30 ohms/sq to approximately 1.5 ohms/sq. This is accomplished by cladding the diffusions and polysilicon gates and emitters with a refractory metal silicide, usually titanium disilicide. Figure 3.27 shows the process flow that is typically used to silicide the MOS gate and source/drains. In this example, an LDD transistor structure is formed using the techniques previously discussed in section 3.4. It is important that the diffusions be driven deep enough to prevent silicide from spiking through the source/drain junctions. An LDD MOSFET structure, just prior to silicidation, is shown in Fig. 3.27(a).

- **POLY GATE PATTERN AND ETCH, REACHTHROUGH IMPLANT TO FORM LIGHTLY DOPED DRAIN EXTENSION**
- **SIDEWALL OXIDE DEP AND ETCH, S/D IMPLANT AND ANNEAL**

- **HF DEGLAZE**
- **SPUTTER DEPOSITION OF TITANIUM**

- **TITANIUM/SILICON REACTION**
- **TITANIUM NITRIDE STRIP**
- **ANNEAL**

Figure 3.27 Fabrication of MOS transistors with low resistivity gates and junctions using the self-aligned TiSi2 process.

The self-aligned silicide process steps include an HF deglaze, titanium deposition, titanium-silicon reaction, titanium nitride strip and silicide anneal. The HF deglaze is done immediately before the 100 nm titanium deposition to ensure a clean interface and promote uniform reaction. A MOS transistor structure, after Ti deposition, is shown in Fig. 3.27(b). The titanium-silicon reaction is performed in a nitrogen ambient, and special precautions are taken to avoid oxygen contamination in the furnace tube, since Ti readily reacts with oxygen to form unwanted titanium oxides. The nitrogen diffuses into and reacts with the Ti over oxide regions to form a stable layer of TiN, which acts as a diffusion barrier. If TiN is not formed during the reaction, silicon will diffuse laterally from the gate and source/drain regions into the titanium over oxide regions, and convert the Ti to silicide. This can result in the formation of a silicide film on top of the sidewall oxide, which will short the gate to the source/drain [3.47]. Thus, the sidewall oxide spacers are performing a dual role. Not only do they mask the high dose source/drain ion implantations from the LDD regions, but they also protect the polysilicon gate sidewalls from silicidation, thereby preventing gate-to-source/drain shorts. Following the silicide reaction, titanium nitride and any unreacted titanium are removed with a standard sulfuric acid/hydrogen peroxide cleanup. An additional anneal is used to reduce the sheet resistance of the silicide to its final value of 1-2 ohms/sq. A MOS transistor structure, after silicidation, is shown in Fig. 3.27(c).

When incorporating the self-aligned titanium silicide into a process flow, it is important to consider the subsequent processing temperatures that will be required. If subsequent heat treatments are not optimized, the active devices can actually exhibit characteristics that are inferior to those achievable without salicidation [3.48, 3.49]. For design rules of 1.5 μm or greater, MOS processes typically include a high temperature step (> 900 °C) to reflow a BPSG interlevel dielectric so as to smooth the surface topography for improved metal step coverage. Unfortunately, the boron in the PMOS source/drains diffuses rapidly into the overlying silicide layer during BPSG reflow (even for the case of rapid thermal processing), causing dopant depletion at the silicide/junction interface and consequently, high contact resistance between the junction and the silicide layer. The problem is illustrated in Figs. 3.28 and 3.29, which respectively show silicided 1 μm gate length PMOS devices that have had either no high temperature backend processing or a 20 s 1150°C flash lamp BPSG reflow. Even though both devices received a high dose S/D implant of 5×10^{15} atoms/cm^2, the device that was processed through the reflow step shows degraded drive current capability due to increased source/drain resistance.

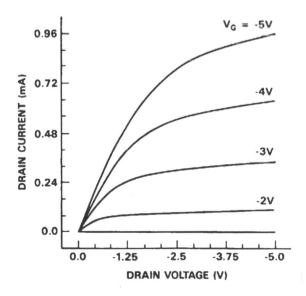

Figure 3.28 Characteristics of a silicided PMOS transistor: W/L=9 μm/1 μm, no high temperature processing after the 800°C silicide anneal.

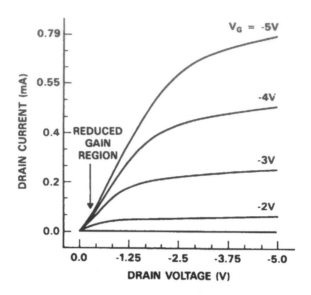

Figure 3.29 Effect of high temperature processing on the performance of silicided PMOS transistors: W/L=9 μm/1 μm, BPSG reflow performed after silicidation using a 20 s, 1150°C flash lamp process.

This series resistance is associated with the formation of a Schottky barrier at both the source and drain junctions. The "kink" at the bottom of the linear region of the device characteristic is caused by a forward-biased Schottky barrier at the drain connection. However, of more importance is the reduced saturated drive current capability that is caused by the reversed-biased Schottky diode in series with the source junction. Many of the advantages of junction silicidation are lost through this mechanism, and a low temperature backend process is required if the full performance advantages of scaled CMOS and BiCMOS processes are to be realized. The necessity of a low temperature backend process (< 800°C after silicidation), is consistent with the general requirements of submicron CMOS and BiCMOS technology, where high temperatures must be eliminated in order to minimize junction depths and maintain doping profiles.

3.6.2 Local Interconnect

As BiCMOS technology is scaled to the submicron regime, the use of the traditional buried contact process to increase the packing density of random logic and SRAM cells becomes less attractive. This is because dopant outdiffusion from the polysilicon that connects gates and emitters to diffusions reduces isolation integrity and adversely affects active device characteristics. These problems limit buried contact design rule scaling, which restricts packing density improvements, as shown in Fig. 3.30. Also, for BiCMOS, phosphorus-doped polysilicon is often used for both the NMOS and PMOS gates as well as the bipolar emitters; this limits the use of buried contacts to the NMOS transistors only.

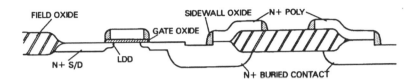

Figure 3.30 Limitations of buried contacts: 1) phosphorus outdiffusion limits device isolation and buried contact-to-gate design rule scaling, 2) the use of N-type polysilicon limits buried contacts in CMOS to NMOS transistors only.

The above limitations make the traditional buried contact process incompatible with VLSI BiCMOS. A technology to replace buried contacts is therefore required to realize the increased packing density made possible through the direct connection of N-type polysilicon gates and emitters to both N+ and P+ diffusions. A local interconnect (LI) technology [3.50] that provides this capability has been realized by utilizing the titanium nitride layer that forms during the self-aligned $TiSi_2$ process described above. The TiN approach requires no extra deposition steps since it utilizes a layer that is normally discarded. Figure 3.31 illustrates how the self-aligned $TiSi_2$ process has been modified to utilize the TiN layer for local interconnections between gates and junctions. After the silicide reaction step, the conductive (~100μ ohm-cm) TiN layer is patterned and etched. Following resist strip, a 800 °C anneal in argon reduces the $TiSi_2$ and TiN resistivities to their final values of 1.0 ohm/sq for $TiSi_2$ and 15 ohm/sq for TiN. This process is capable of realizing LI linewidths as narrow as 0.8 μm with contact resistances to diffusions clad with $TiSi_2$ as low as 2×10^{-8} ohm-cm^2.

Figure 3.32 is a CMOS process cross-section showing how the TiN layer of LI is used to implement full buried contacts. The TiN layer provides direct connection of the drains of the first inverter stage, to the common gate of the second stage without area consuming contacts and metal straps. By implementing LI, the number of contacts is greatly reduced and those contacts that are required can be more readily shared. LI also allows junctions to be "extended" over the isolation regions so that minimum geometry junctions can be realized to reduce capacitance and hence increase circuit speed [3.50]. The use of LI results in NMOS, PMOS and bipolar base junction capacitances being reduced by up to 50% since the diffusions no longer have to accomodate contacts. Results from CMOS inverter chains show up to a 15% improvement in propagation delay when LI is used to strap the drains of one inverter to the input of the next, compared to the traditional approach where contacts and metal interconnect are used [3.5].

An example of how LI technology can be used to increase circuit packing density is shown in Fig. 3.33. In this example, LI is used to reduce the size of a six-transistor CMOS SRAM cell by replacing the area-consuming contacts and metal straps used for the cross-coupling function. Using 0.8 μm design rules, the conventional double-level-metal (DLM) cell layout [3.51] is reduced from approximately 160 μm^2 to 118 μm^2 when using LI, representing a 25% reduction in area.

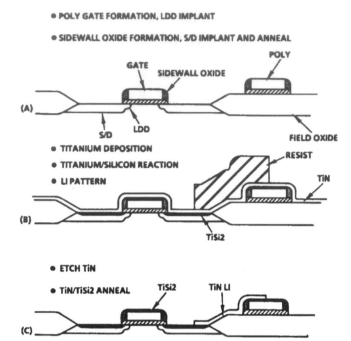

Figure 3.31 Modification of the self-aligned titanium silicide process to realize a level of local interconnect using titanium nitride.

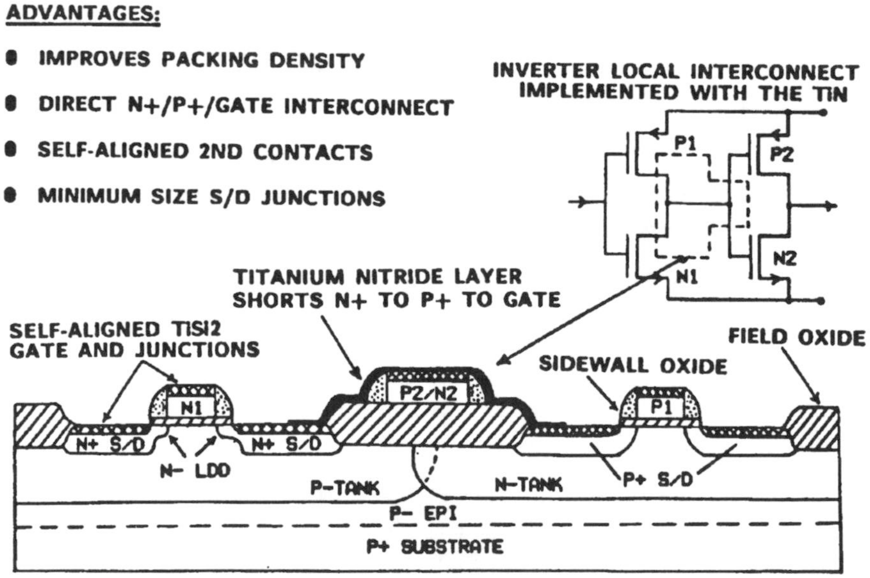

ADVANTAGES:

● **IMPROVES PACKING DENSITY**

● **DIRECT N+/P+/GATE INTERCONNECT**

● **SELF-ALIGNED 2ND CONTACTS**

● **MINIMUM SIZE S/D JUNCTIONS**

Figure 3.32 Process cross-section of a CMOS inverter showing how the TiN local interconnect level is used to connect both N+ and P+ diffusions to N-type polysilicon gates.

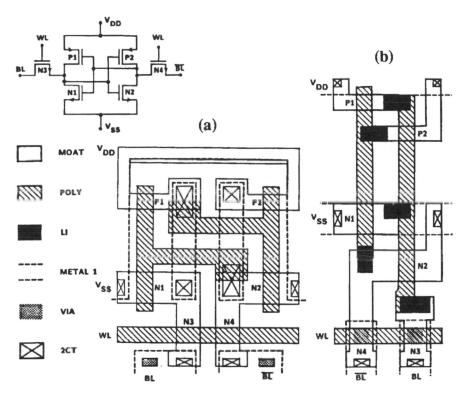

Figure 3.33 Example of how LI technology can be used to decrease the size of a six-transistor CMOS SRAM cell by approximately 25%: (a) Conventional layout, area=160 μm² (b) Cell cross-coupling implemented with TiN LI, area=118 μm².

3.6.3 Planarization and Metallization

A key requirement of submicron BiCMOS technology is a high density, high reliability multilevel metallization system. As dimensions approach the submicron regime, the reduced depth of focus associated with higher resolution optical lithography tools calls for greatly improved planarization of the oxide layers used between the metal levels. In addition, the techniques needed to improve planarization must be compatible with the lower temperatures (< 800°C) needed to maintain critical doping profiles, shallow junction depths and low silicide-to-diffusion contact resistance. The metallization systems must be capable of permitting the use of high current densities without elctromigration induced reliability problems. In addition, the interconnect techniques and materials must provide reliable metal step coverage into near vertical submicron contact holes and vias that will typically have greater than a 1:1 depth to diameter aspect ratio. (Contact holes are defined as the holes that are etched in the first interlevel oxide allowing connections to be made between the silicon substrate and the first level of metal. Vias are defined as the holes that are etched in the upper interlevel oxides allowing connections to be made between metal levels).

The above requirements have resulted in revolutionary changes in submicron interconnect and planarization techology compared to those commonly employed for the 1.5-2 μm class of design rules [3.52]. The more traditional approaches of high temperature (900°C or greater) PSG or BPSG reflow processes designed to improve metal coverage over steps and into contacts have given way to low temperature planarization techniques such as resist etchback planarization [3.53], spin-on glass and oxide deposition followed by etchback. The method of using resist etchback to obtain adequate planarization at 0.8 μm design rules is shown in Fig. 3.34. This technique consists of first depositing a conformal oxide over the underlying interconnect level and then coating the wafer with photoresist, as shown in Fig. 3.34(a). The photoresist has a planarization range of several microns, and this profile is then etched anisotropically into the underlying oxide using a plasma etch that has approximately a 1:1 etch ratio between oxide and photoresist. During the plasma etch, the photoresist clears first in those regions with the highest underlying topography and therefore starts to etch the oxide in those locations while photoresist still protects regions of lower topography. The etch is stopped prior to clearing the oxide over the highest topography regions, and then any remaining photoresist is stripped, resulting in the structure shown in Fig. 3.34(b). A layer of oxide is then deposited to increase the interlevel oxide to its final value, typically 0.8 μm over the highest topography regions. After via formation (not shown), the upper level of interconnect is deposited and patterned, as shown in Fig. 3.34(c). This approach results in a well-planarized surface that permits the patterning of submicron vias and metal interconnections.

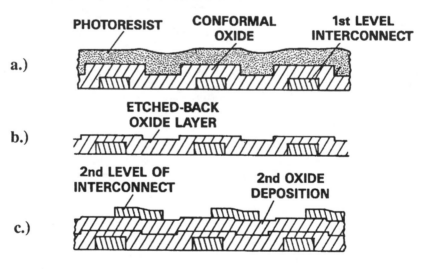

Figure 3.34 Planarization of interlevel oxides using the technique of resist-etch-back: (a) cross-section of lower level of interconnect after interlevel oxide deposition and photoresist coating, (b) cross-section after etchback using an etch with 1:1 selectivity between oxide and photoresist, (c) cross-section following deposition of the second layer of oxide, via formation (not shown) and definition of the upper level of interconnect.

Another feature of advanced submicron BiCMOS interconnect technology is the introduction of contacts and vias with vertical walls, designed to replace their sloped-wall counterparts for improved packing density. This, in turn, is leading to new metallization systems, such as CVD tungsten, which are needed to completely fill the vertical contacts and vias and avoid the step coverage and electromigration problems associated with traditional aluminum-based metallization systems. For example, by depositing approximately 0.8 μm of blanket CVD tungsten on a thin composite Ti:W seed layer, a CVD tungsten layer simultaneously plugs contacts while serving as the first level of metal interconnect in a multilevel-metal (MLM) integrated circuit [3.54]. This process results in a high reliability metal system that has a resistivity of approximately 7 μohm-cm. Blanket CVD tungsten is used for contact "plugging" rather than selective tungsten (a process where the deposition conditions are adjusted such that the tungsten forms only in the contacts and not on the multilevel oxide) because it avoids the risk of damaging the diffusions through silicon consumption. Intermediate levels of contact and via filling in a multilevel-metal integrated circuit may also be fabricated in the same way using blanket-deposited tungsten. In contrast, the top level of metal is often a different material, such as aluminum-copper, due to the lower resistivity requirements of power busses. In this case, the selective tungsten deposition process can be used to plug the top level via, followed by the deposition of a composite layer of Ti:W/Al-Cu that has a sheet resistance of typically 40 millohms/sq as compared with approximately 80 millohms/sq for the CVD tungsten levels. The introduction of Ti:W under the Al-Cu helps improve the electromigration properties of this top layer of metal.

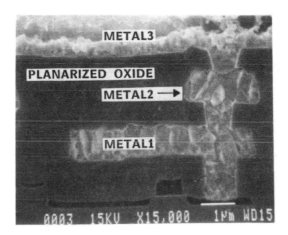

Figure 3.35 Use of CVD tungsten to permit stacked contacts and vias in a triple level metal process.

These advanced metallization and planarization techniques also permit vias to be stacked on top of lower-level vias and contacts. As shown in the triple-level-metal example of Fig. 3.35, this capability permits the top level of metallization to directly contact the silicon substrate through stacked contact/vias, circumventing the need for lower-level interconnects and greatly reducing circuit layout area. Note that the example interconnect system is also capable of achieving non-overlapped contacts and vias, in contrast to previous interconnect systems which have required contacts/vias to be "nested" inside overlapped metal to ensure adequate coverage under worst case misalignment conditions. Nesting has the disadvantage of increasing the minimum metal pitch including contacts/vias (referred to as the "working" pitch) as compared to the pitch that can be used without contacts and vias. With the advent of "plugged" contacts and vias, it is now possible to eliminate the previously required overlaps and make the "working" pitch equal to the minimum pitch.

For BiCMOS applications that incorporate a large number of ECL-based circuits, metal bus voltage drops can become a critical functionality and performance issue. For this type of ECL-intensive BiCMOS design, the resistivity of the lower levels of metal interconnects has to be reduced to typically 3 μohm-cm, which is below the 7 μohm-cm achievable with CVD tungsten. To satisfy the ECL-intensive design requirments, alternative metallization systems such as gold and copper are under investigation [3.55,3.56]. Such systems would still use tungsten for filling the contacts and vias, but the lower resistivity metals would be used for interconnections.

3.7　Example Submicrometer BiCMOS Process Flow for 5V Digital Applications

In approaching BiCMOS from a CMOS perspective, it is advantageous to merge the bipolar elements into a baseline CMOS process without altering the electrical characteristics of the CMOS device. Using this approach, the BiCMOS process flow becomes transparent to all CMOS circuits designed within the original CMOS-only guidelines, improving the cost effectiveness of the mixed technology. For example, standard cells designed for the CMOS-only process can be retrofitted with ECL I/O without changing the internal CMOS core circuit for improved system performance with minimal design investment.

In keeping with this approach, this section discusses an example 0.8 μm BiCMOS process flow [3.3], emphasizing performance, reliability, process simplicity and compatibility with a same generation 0.8 μm baseline CMOS technology [3.5]. Process commonality between the BiCMOS and CMOS flows has been adopted wherever possible, and the extra process steps required for bipolar transistor fabrication, while involving dramatic changes to the isolation structure, have been introduced with only minimal changes to the actual CMOS transistor fabrication sequence. Table 3.1 shows how the bipolar processing steps have been merged into the baseline CMOS process flow. Key steps in the example 0.8 μm BiCMOS process flow are illustrated in Figures 3.36 through 3.47 and are described below:

Table 3.1 Process flow showing how the bipolar processing steps have been merged into a baseline CMOS process.

CMOS BASELINE FLOW	CHANGES FOR BIPOLAR
P+ SUBSTRATE	**P- SUBSTRATE**
	BURIED N+/P-LAYER
P- EPI (5μm)	**INTRINSIC DOPED EPI (1.3μm)**
N-WELL/P-WELL	
TANK DRIVE	**REDUCED DRIVE TIME**
POLY BUFFER LOCOS	**HIGH PRESSURE OXIDATION**
	DEEP COLLECTOR/N+ RESISTOR
	BASE/P RESISTOR
VT IMPLANT	
GATE OXIDATION (200 A)	
	THIN POLY DEPOSITION
	EMITTER PATTERN/ETCH
POLY DEPOSITION/DOPING	**IMPLANT POLY EMITTER DOPING**
PATTERN/ETCH POLY	
LDD PATTERN/IMPLANT	
SWO DEPOSITION/ETCH	
PATTERN/IMPLANT N+/P+ S/D	
ANNEAL S/D	**ANNEAL OPTIMIZED FOR EMITTER**
TiSi2 FORMATION	
PATTERN/ETCH LI	
STANDARD DLM FLOW	

3.7.1 Starting Wafer and N+ Buried Layer Formation-

A lightly-doped (~10 ohm-cm) P-type substrate is used for the starting wafer. The buried N+ layer is formed by implanting antimony into windows etched in a thick oxide covering the substrate. Afterwards, a high temperature anneal (1250°C) is used to diffuse the antimony and remove defects. During this anneal, an oxide is grown in the buried N+ windows to provide a silicon step for alignment of subsequent levels. After stripping all the oxide, a patterned boron punchthrough implant is used to maintain adequate breakdown between buried N+ regions. An alternative self-aligned approach has been reported [3.1] which utilizes a nitride mask to prevent oxidation of P-type regions following the buried N+ implant. As shown in Fig. 3.36, the nitride and pad oxide sandwich is patterned and etched before implanting antimony. During the buried N+ anneal, an oxide is grown over the implanted region just as in the traditional bipolar approach. However, the nitride mask can now be selectively removed, leaving a thick oxide over the buried N+ regions and a thin oxide over the P-type regions. The thick oxide serves as a blocking mask for the self-aligned boron buried P-layer implant, as illustrated in Fig. 3.37.

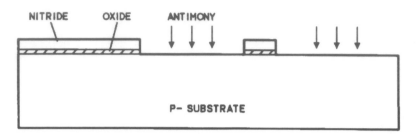

Figure 3.36 Device cross-section for BiCMOS process flow showing buried N+ implant.

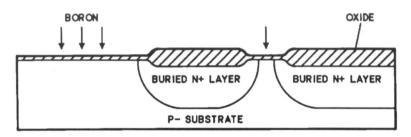

Figure 3.37 Device cross-section for BiCMOS process flow showing buried P implant self-aligned to buried N+.

3.7.2 Epitaxial Layer Deposition

All oxide is removed from the surface prior to epitaxy, and a short HCl etch is done in situ to remove surface defects before growing a thin (1-1.5 μm) N-epitaxial layer with intrinsic doping (> 15 ohm-cm resistivity).

3.7.3 Twin-well Formation

The wafer is oxidized and capped with a nitride layer, which is selectively removed from the N-well regions by a patterning step. The N-well dopant is then implanted in a similar way as shown in Fig. 3.36 except that the photoresist mask is retained to prevent the double (shallow and deep) phosphorus implant from entering the P-well region. After resist removal, a thick (~350 nm) oxide is grown over the N-well regions and the nitride is stripped from the P-wells. The subsequent P-well implant is self-aligned to the N-well edge; the N-well oxide prevents counterdoping of the N-well as shown in Fig. 3.38. A relatively short (~150-250 min) well drive is performed at 1000°C with the oxide cap in place.

3.7.4 Active Region and Channel Stop Formation

All oxide is removed from the surface and a thin (~10 nm) pad oxide is grown and capped by an additional polysilicon buffer pad (~50 nm thick) and thick nitride (~240 nm). After patterning active device regions, anisotropic plasma etches are used to remove the nitride and polysilicon layers from the field isolation regions. A blanket boron channel stop is implanted into the field regions with the resist in place, as shown in Fig. 3.39. High pressure oxidation at 10 atm and 975°C is used to grow 800 nm of field oxide so as to minimize buried layer diffusion.

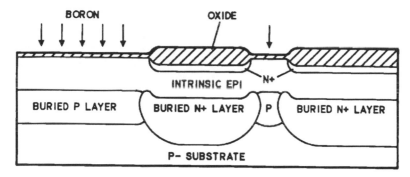

Figure 3.38 Device cross-section for BiCMOS process flow showing P-well implant self-aligned to oxide-masked N-well.

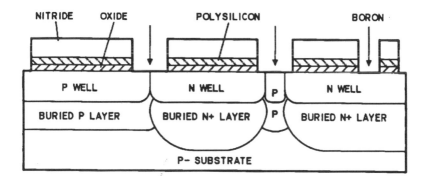

Figure 3.39 Device cross-section for BiCMOS process flow showing channel stop implant after etch of field isolation regions.

3.7.5 Deep N+ Collector and Base Formation

Following the removal of the nitride oxidation mask and underlying polysilicon and oxide buffer layers, a thin (~20 nm) pregate oxide is grown in the active device regions. Deep collector regions are patterned and implanted with phosphorus, as shown in Fig. 3.40. A double implant (shallow/deep) may be used to provide a lower resistance path to the buried N+ subcollector. After resist removal, combined P-type base and resistor regions are patterned and implanted as shown in Fig. 3.41. A thicker base oxide is formed over the base region to limit oxidation-enhanced diffusion during the subsequent gate oxidation and also provide lower capacitance between the base and polysilicon emitter. This oxide can be selectively grown over the base before the implant (so the thermal cycle does not affect the base) or deposited and selectively removed from non-base/resistor regions after base implant. An additional boron dose is then implanted for CMOS threshold adjust; the thicker base oxide prevents this implant from affecting the base region.

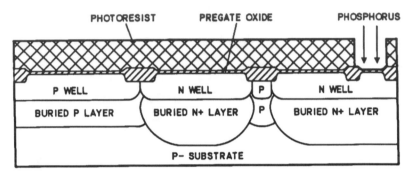

Figure 3.40 Device cross-section for BiCMOS process flow illustrating patterned deep collector implant.

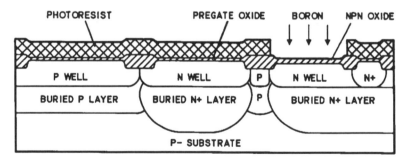

Figure 3.41 Device cross-section for BiCMOS process flow showing patterned base and P-type resistor implant.

3.7.6 Polysilicon Emitter Formation

The active emitter is formed using a split polysilicon process [57]. First, the pregate oxide is removed and a thin (20 nm) gate oxide is grown and capped by an initial polysilicon layer. The active emitter window is then patterned and anisotropically etched through the polysilicon and underlying oxide. A second polysilicon layer is deposited and forms a contact with the base through the emitter window. Finally, a blanket phosphorus implant is used to introduce the polysilicon dopant that will ultimately form the diffused emitter junction (Fig. 3.42).

3.7.7 Gate and LDD Formation

Next, the polysilicon is patterned and anisotropically etched to define both CMOS gates and polysilicon emitter regions. A phosphorus implant is selectively implanted to form a shallow, N-type LDD region which is self-aligned to the NMOS gate edge, as shown in Fig. 3.43 and previously discussed in Section 3.4.

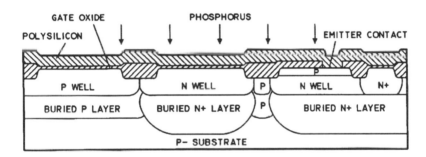

Figure 3.42 Device cross-section for BiCMOS process flow after emitter formation.

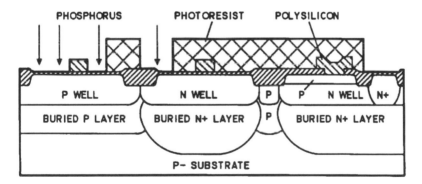

Figure 3.43 Device cross-section for BiCMOS process flow showing reachthrough pattern and implant.

3.7.8 Sidewall Oxide and Final Junction Formation

Before introducing the heavily doped source/drain implants, a sidewall oxide of approximately 300 nm thick is formed by depositing a conformal oxide layer (such as a tetraethyl orthosilicate/TEOS) and using an anisotropic etchback. N+ and P+ source/drains and respective well and substrate contacts are then patterned and implanted as illustrated in Figs. 3.44 and 3.45. Note that the P+ source/drain pattern also opens the extrinsic base region of the bipolar transistor. A 900°C anneal simultaneously forms the diffused emitter, extrinsic base and N+/P+ source/drain junctions.

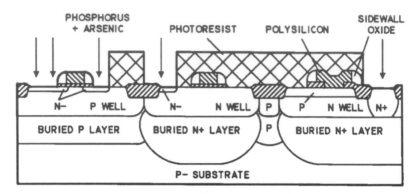

Figure 3.44 Device cross-section for BiCMOS process flow showing N+ source/drain pattern and implant.

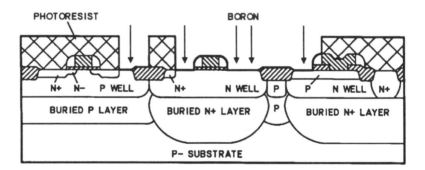

Figure 3.45 Device cross-section for BiCMOS process flow showing P+ source/drain pattern and implant.

3.7.9 Silicide and Local Interconnect Process

Silicidation of the polysilicon and exposed active regions is achieved in conjunction with local interconnect patterning, as discussed in Section 3.6. The $TiSi_2$ provides a low sheet resistance (~1.5 ohms/sq) for the source/drain and extrinsic base regions, as well as for the polysilicon emitter, gate and interconnect. Patterned TiN straps (~15 ohms/sq) are used as low capacitance source/drain extensions and for local interconnects as shown in Fig. 3.46.

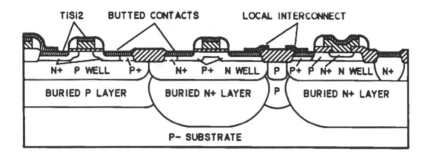

Figure 3.46 Device cross-section for BiCMOS process flow illustrating silicide and local interconnect structure.

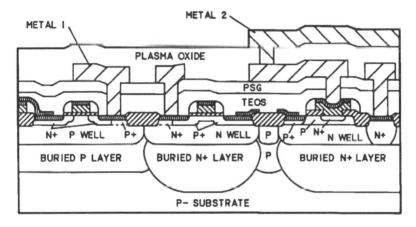

Figure 3.47 Device cross-section for BiCMOS process flow following completion of double-level metal interconnect.

3.7.10 Multilevel Metal (MLM) Processing

After silicide and LI formation, a conformal oxide is deposited and planarized using the resist etchback technique discussed in Section 3.6. A PSG layer is then deposited and activated to act as a sodium barrier. Contacts are patterned and etched through the interlevel oxide, and the contacts are subsequently filled using CVD tungsten. The conformal metal deposition produces very reliable contacts, since problems with poor metal step coverage into the deep contact holes are avoided. Additional levels of metal interconnect can be formed by sequentially performing: (1) a low temperature oxide deposition, (2) resist etchback planarization, (3) via pattern and etch, (4) CVD tungsten deposition and (5) metal leads pattern and etch. For the top leads, an aluminum-copper cap is used to provide lower sheet resistance and ease of bonding. As a final example, Fig. 3.47 illustrates a completed double-level metal (DLM) device cross-section.

In summary, an example 0.8 μm BiCMOS process flow has been described which realizes high performance NPN bipolar devices through the use of polysilicon emitters fabricated with a split polysilicon process that offers strong compatibility with a baseline CMOS process. Four additional mask levels are required to merge the bipolar devices (including a P-type resistor) into the baseline CMOS process flow, providing an efficient and cost effective way of implementing the mixed BiCMOS technology.

3.8 Analog BiCMOS Process Technology

3.8.1 Analog BiCMOS Evolution

BiCMOS is not a new technology in the fabrication of analog designs. Analog designers have for some time taken advantage of the superior transconductance, bandwidth, matching, and noise characteristics of bipolar transistors, while adding the high input impedance and switching capabilities offered by MOS devices. In recent years, however, the complexity of analog BiCMOS processes has increased markedly with the advent of Linear ASIC cell strategies, where high performance analog and digital functions are combined on a single chip. This trend toward higher levels of analog/digital integration is likely to continue. Unlike most digital functions, however, analog functions can be extremely diverse (operating voltage, power handling, etc.), and can require significantly different process technologies to be implemented effectively. As a result of this diversity, two types of analog BiCMOS technologies have emerged: 1) low-to-medium voltage (10-30 V), CMOS-oriented processes aimed at high performance and packing density, and 2) high voltage/power (>30 V, >1 A), bipolar-oriented processes intended for power applications, which often include DMOS components.

The transition from bipolar to BiCMOS took place in increments, with the addition of PMOS devices onto an existing standard buried collector (SBC) process [3.58], which was later upgraded to include full CMOS capability [3.59]. The MOS components were used mainly to provide high impedance inputs for op-amps and CMOS output inverters. The above processes utilized metal-gate MOS technology due to compatibility with existing bipolar flows.

Threshold voltage instabilities associated with metal-gate CMOS technologies [3.60] and erratic control of bipolar NPN gains led to a gradual shift toward silicon-gate CMOS and BiCMOS processes [3.61,3.62]. The use of N+ polysilicon gates preserved the cleanliness of the gate oxide while avoiding the polarization shifts caused by the PSG (phosphosilicate glass) sodium gettering layer used in metal-gate technology [3.63]. Also, by growing the gate oxide prior to the emitter process, improvements in bipolar NPN h_{FE} control could be realized.

In recent years, the demand for higher levels of functional integration coupled with reductions in system power supply levels has sparked great interest in low-voltage, high-performance analog BiCMOS technologies using sophisticated standard cell and gate array methodologies. Due to the increased demand for digital logic integration, a move toward CMOS-oriented, and CMOS-flavored bipolar-oriented, BiCMOS processes has occured [3.64-3.67]. These approaches are well suited for analog VLSI, offering the superior analog performance of scaled bipolar coupled with the analog stability and high-density logic capability afforded by silicon-gate CMOS.

3.8.2 Analog BiCMOS Process Design Considerations

The special requirements of analog functions create many new and different considerations for analog process design than are encountered in digital processes. Often, these considerations are in conflict, forcing process compromises to be made based on cost-effectiveness and specific product requirements. A sampling of these considerations is presented below.

3.8.2.1 Operating Voltage

Most high-performance analog BiCMOS designs require an operating voltage of 10-15 volts. This is due to the need for high signal-to-noise ratios and the large use of cascoding in analog designs. Both bipolar and CMOS devices used in the analog sections must be able to reliably withstand this voltage. Still other analog applications, such as automotive circuits, generally require 15-30 volt operation or greater.

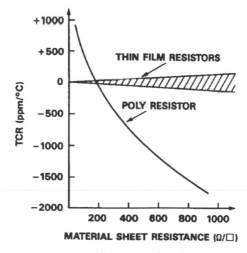

Figure 3.48 Resistor temperature coefficients as a function of sheet resistance for different resistor materials. Phosphorus, boron, and arsenic have been used to dope the polysilicon resistors.

3.8.2.2 Passive Components

Any analog process must provide high quality resistors and capacitors with low temperature and voltage coefficients. High-value polysilicon resistors are preferable to silicon resistors, which suffer from depletion effects. Thin film resistors are often used in high precision applications [3.68]. Figure 3.48 shows the temperature coefficient for polysilicon and thin film resistors [3.69] as a function of sheet resistance. Polysilicon-polysilicon MOS capacitors are used to reduce parasitic effects associated with polysilicon-silicon capacitors. The polysilicon should be heavily doped to minimize depletion effects in the capacitor electrodes.

3.8.2.3 Bipolar Components

The NPN transistor is the dominant device in analog design. An isolated structure with low collector resistance (R_{cs}) and high Early voltage (V_A) is needed for high current operation and increased output resistance respectively. Low R_{cs} can be accomplished through the use of an N+ buried layer structure, while higher base Gummel numbers increase V_A. However, the need for thicker epitaxial layers (due to higher voltage operation) and higher base Gummel numbers are inconsistent with high bandwidth (f_T) transistor design [3.70]. Additionally, isolated PNP transistors are necessary in analog designs. Generally, the lateral PNP device is the only isolated PNP and is slow (f_T = 5-10 MHz) due to the use of wide base regions to avoid punchthrough of the emitter/collector areas. While p-type mosfets can often replace PNPs, there is a need for faster isolated PNP structures, but the cost and complexity of any such process module are serious obstacles to their implementation.

3.8.2.4 CMOS Components

Analog CMOS devices generally do not scale well due to reduced output resistance caused by channel length modulation and drain-induced barrier lowering (DIBL) effects [3.71]. This typically results in minimum channel lengths of 2-3 μm. Thus, lightly doped wells are preferred in analog CMOS devices for reduced capacitance. In addition, since analog designs frequently use dual (+/-) supply voltages, isolated CMOS devices are desirable. This allows the digital logic NMOS devices to be shielded from the substrate bias without the need for level-shifting techniques, which increase delays and degrade standard cell compatibility.

3.8.2.5 Component Matching

Analog designs require precise matching for both active and passive components, as well as for component temperature coefficients. In many cases the matching between components is more critical than the absolute value of the components. Matching is a strong function of device size [3.72], necessitating the design of large component geometries and tight patterning and etching control in the fabrication area. Closely associated with the requirement for matching is the desire to trim circuit parameters after completion of the process. Polysilicon and metal fuses, zener zaps, and EEPROMs are all means of trimming analog circuits. While circuit trimming at wafer probe is usually acceptable, the ability to trim circuits in packaged form is desirable to offset assembly-induced shifts.

3.8.2.6 Noise

Noise is a major consideration in analog designs. Bipolar devices generally exhibit much better noise performance than MOS devices [3.73]. High intrinsic base sheet resistances can adversely affect bipolar noise by increasing R_B through higher pinch resistance. The high 1/f noise in MOS devices can be reduced through the use of thinner gate oxides and/or larger gate areas. The latter approach is preferable, since thinner gate oxides reduce the maximum operating voltage.

3.8.2.7 Reliability

The higher voltages used in analog designs present potential charge-spreading problems whereby charge from high-voltage sections can spread to low-voltage regions, resulting in deleterious effects. Higher thick field oxide thresholds, above the maximum operating voltage, are required since field-plating alternatives are typically area-intensive. Additionally, since the analog CMOS devices are exposed to higher voltages, they must be protected against hot-carrier problems, which can degrade device performance over time. Figure 3.49 shows the impact of gate oxide, channel length, and operating voltage on NMOS substrate current, an indicator of hot-electron susceptibility. Latchup is another concern in analog BiCMOS processes due to the high substrate currents common in many analog designs. An improved substrate contact, using either deep P+ diffusions, P+ guardrings, or backside contact, may be needed depending on the particular application. Also, special layout rules may be needed to reduce the onset of substrate de-biasing.

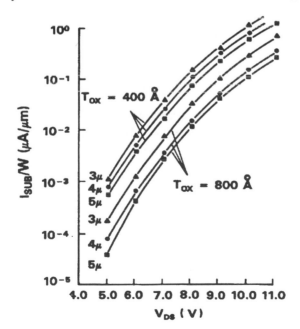

Figure 3.49 Normalized NMOS maximum substrate current as a function of operating voltage for various process parameters.

3.8.3 Analog BiCMOS Process Integration Discussion

A primary decision in BiCMOS technology development is which starting technology (bipolar or CMOS) to use. Bipolar-oriented processes suffer from poor packing densities due to the use of thicker epitaxial layers and the need for deep P+ isolation. N-well CMOS processes, however, allow the NPN collector region to be self-isolating with the P-epitaxial layer acting as the isolation region. Since N-well CMOS processes are commonly used in analog and digital applications, this choice of starting process can minimize the amount of device recharacterization needed after merging the bipolar components. Hence, the following BiCMOS process integration discussion focus on N-well CMOS-oriented strategies.

The necessary bipolar process steps should be integrated into the existing CMOS process without perturbing the original CMOS parameters (both electrical I-V data as well as layout rules). This affords the opportunity to later import existing CMOS standard cells into any new cell library based on the BiCMOS process. The major process decisions that must be made when integrating the bipolar steps are: 1) what type of N+ buried layer process is needed, 2) what P- epitaxial layer thickness is needed to support the bipolar operating voltage, 3) where to add the deep N+ collector diffusion, 4) whether to use the source/drain regions as emitter and base regions, and 5) where to integrate the separate emitter and base operations if they are deemed necessary.

The N+ buried layer process is typically the first module in the BiCMOS process and, as such, the heat-cycling associated with this operation does not affect existing CMOS device profiles. A highly doped and shallow N+ region is desired, resulting in low NPN R_{cs} as well as high packing density. Sheet resistances of 20 ohms/sq and junction depths of 2.5 μm are typical. The possibility of N-layer autodoping in the P-epitaxial layer dictates the use of antimony.

The choice of epitaxial layer resistivity and thickness impacts both bipolar and CMOS device charcteristics. By choosing the resistivity to be the same as the P-substrate resistivity in the CMOS-only process and retaining the same N-well process, the CMOS device performance can be maintained. Determining the necessary epitaxial layer thickness, however, is somewhat more complicated and is dependent on such issues as: 1) N-well junction depth, 2) N-well doping level, 3) N+ buried layer up-diffusion, 4) the amount of P-epitaxial layer removed through subsequent oxidations, 5) maximum allowable NPN h_{FE}, and 6) the maximum operating voltage. The following illustrates the process involved in choosing the P-epitaxial layer thickness.

In order to form low-resistance NPN collectors, the N-well must extend down to contact the N+ buried layer. Hence, there is a maximum epitaxial layer thickness for a given N-well depth which is shown in Fig. 3.50. For a heavily doped, shallow N+ buried layer process the amount of up-diffusion can be on the order of 80-100 percent of the N-well junction depth.

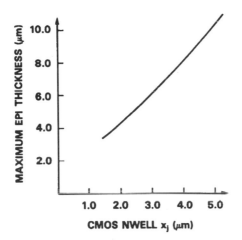

Figure 3.50 Maximum BiCMOS P-epitaxial layer thickness as a function of CMOS N-well diffusion depth. Approximately 1.0 micron of the P-epitaxial layer was removed through oxidations in this example.

In most analog BiCMOS processes, the NPN operating voltage is limited by the BV_{CEO} breakdown mechanism. This is related to the epitaxial layer thickness and NPN h_{FE} as shown in the following empirical equation [3.74]:

$$BV_{CEO} = \frac{BV_{CBO}(plane)}{\lceil h_{FE} \rceil^{1/n}} \qquad (3.8.1)$$

where $BV_{CBO}(plane)$ is the collector-base breakdown voltage and the parameter n has a typical value of 3-6.

Here, $BV_{CBO}(plane)$ refers to the noncurvature-limited collector-base breakdown voltage along the planar (i.e. flat) portion of the base region, directly under the emitter region. (This portion of the collector-base junction is important in this calculation since the majority of the collector current flows across this region of the junction). It should be noted that generally the actual BV_{CBO} of the device is lower than $BV_{CBO}(plane)$ due to the curvature and channel stop doping effects. Referring again to equation 3.8.1, it can be seen that once hfe and BVceo are set by the design requirements, the necessary $BV_{CBO}(plane)$ can be calculated. For example, a hypothetical process requiring a maximum operating voltage of 15V with a maximum hfe of 150 would need a minimum $BV_{CBO}(plane)$ of 53V, using n=4.

The epitaxial layer thickness is usually the main variable in determining $BV_{CBO}(plane)$, since the N-well and base processes are set by other factors such as CMOS compatibility. As the epitaxial layer thickness is reduced from its maximum, the vertical distance between the base and buried collector diffusion is reduced, lowering $BV_{CBO}(plane)$. Figure 3.51 shows a typical NPN doping profile, while Fig. 3.52 illustrates the impact of epitaxial layer thickness on $BV_{CBO}(plane)$ for a particular N-well process. Using Fig. 3.52, a minimum epitaxial layer thickness of 7.0 μm is found for the above example where the minimum $BV_{CBO}(plane)$ must be greater than 50 volts.

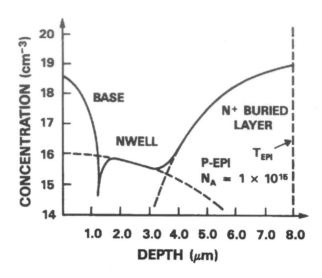

Figure 3.51 BiCMOS NPN doping profiles for BV_{CBO}(plane) calculations. The solid line shows the net concentration while the dotted lines show the original profiles.

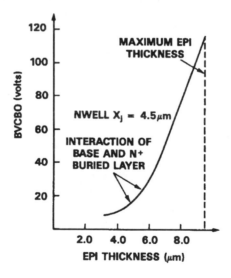

Figure 3.52 BV_{CBO}(plane) as a function of P-epitaxial layer thickness for a 4.5 μm N-well process.

High-current NPN transistors, common in analog designs, often need a deep N+ collector diffusion to lower the vertical collector resistance of the lightly doped N-well. However, this is a deep diffusion (generally as deep as the epitaxial layer thickness) and requires high temperature processing. As such, the diffusion is normally done before the channel stop module, so the additional heat does not impact CMOS device performance.

Efforts to use source/drain regions for bipolar emitter and base regions [3.75] have shown reduced MOS transconductance through increased source/drain sheet resistance. Additionally, the deeper emitter/base junction depths normally required for analog devices are inconsistent with high density CMOS logic needs. Hence, it is generally best to independently optimize both the CMOS and bipolar devices, although this increases the process mask count. The base diffusion, which can involve substantial heat cycles, is often implemented before the channel stop implants, although this reduces bipolar self-alignment. The emitter operation is normally performed near the end of the process so as to minimize process variability, which can result in poor h_{FE} and f_T control.

3.8.4 Sample Analog BiCMOS Process

The process sequence described in this section is meant to illustrate the general methodology used to fabricate a CMOS-oriented analog BiCMOS process.

3.8.4.1 Starting Wafer and N+ Buried Layer Formation

The starting wafer is a P-type (100) oriented substrate. The wafer is oxidized and the N+ buried layer regions are patterned and formed by means of an antimony implant and a subsequent diffusion step (Fig. 3.53).

3.8.4.2 P- Epitaxial Layer Deposition

Before deposition of the epitaxial layer, all oxide is stripped from the wafer surface. A short HCl etch is done to improve the surface quality and a P-type epitaxial layer is deposited. The resistivity and thickness are determined as discussed in section 3.8.3.

3.8.4.3 N-well and N+ Collector Formation

The wafer is oxidized and N-well regions are exposed, implanted, and partially diffused. The N+ collector areas are then formed, with the N+ collector heat cycle completing the N-well drive (Fig. 3.54).

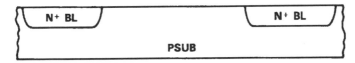

Figure 3.53 Formation of the N+ buried layer regions.

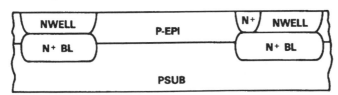

Figure 3.54 Formation of the N-well and N+ collector regions following the P-epitaxial layer deposition.

3.8.4.4 Base Formation

The oxide is removed from the wafer and a thin pad oxide is grown. The base regions are patterned and ion implanted using boron. Next, the implant is diffused in an inert ambient, to anneal the silicon for improved bipolar performance (Fig. 3.55).

3.8.4.5 Active Region and Channel Stop Formation

This process uses standard LOCOS techniques. An LPCVD nitride film is deposited on the wafer, a active region pattern is defined, and the nitride is plasma etched. A blanket phosphorus channel stop implant is performed to raise the PMOS thick field oxide threshold voltage. Subsequently, a second layer of resist is patterned, covering only the N-well regions. A boron channel stop implant is performed in the P-type regions, raising the NMOS thick field oxide threshold voltage. The photoresist layers are removed and a thick field oxide (~1.0 μm) is grown in the nonactive regions (Fig. 3.56).

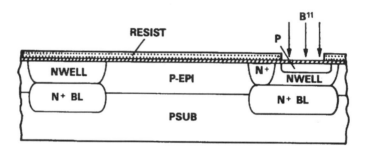

Figure 3.55 Definition of the base regions.

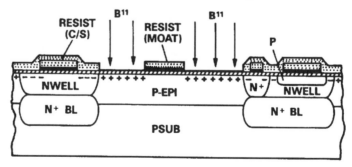

Figure 3.56 Definition of the active regions and subsequent channel stop implants.

3.8.4.6 1st Polysilicon Layer Formation

After removal of the nitride oxidation mask and the underlying pad oxide, a thin gate oxide (350-500 nm) is grown, typically in an O_2/TCA ambient, and a threshold adjust implant is performed. LPCVD polysilicon is then deposited, doped N+, patterned, and plasma etched to form the MOS gate electrodes and the bottom plates of capacitors (Fig. 3.57).

3.8.4.7 Capacitor Dielectric and 2nd Polysilicon Layer Formation

A capacitor interlevel dielectric (ILD) is formed, using either oxide or an oxide/nitride combination, with an effective oxide thickness in the range of 30-100 nm. The second LPCVD polysilicon layer is deposited and doped to form high-sheet resistors with low temperature coefficients. A mask is used to protect the

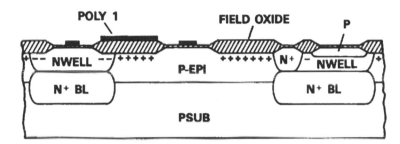

Figure 3.57 Deposition and definition of the 1st polysilicon layer for CMOS gates and MOS capacitors.

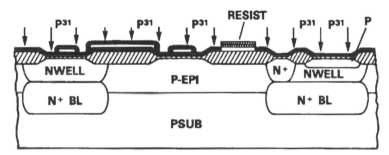

Figure 3.58 Deposition and definition of the 2nd polysilicon layer for resistors and capacitor plates, showing the resist block over the resistor regions during the high-dose capacitor plate implant.

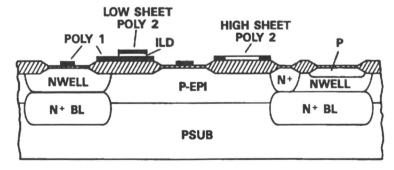

Figure 3.59 Formation of the 2nd polysilicon layer for resistors and capacitor plates.

resistors while the remainder of the polysilicon is doped N+ to form the capacitor top plates and resistor contacts (Fig. 3.58). The polysilicon is then patterned and etched using a uniform etch process to provide good resistor and capacitor matching (Fig. 3.59).

resistor while the remainder of the polysilicon is doped N+ to form the capacitor top plates and resistor contacts (Fig. 3.58). The polysilicon is then patterned and etched using a uniform etch process to provide good resistor and capacitor matching (Fig. 3.59).

3.8.4.8 NMOS LDD Formation

A light phosphorus dose is selectively implanted into the NMOS transistor areas to form the LDD structure. A conformal layer of LPCVD oxide is deposited and anisotropically etched to form oxide sidewall spacers on the gate electrodes (Fig. 3.60).

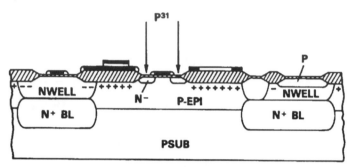

Figure 3.60 Definition of LDD NMOS source/drain regions and formation of the oxide sidewall spacers.

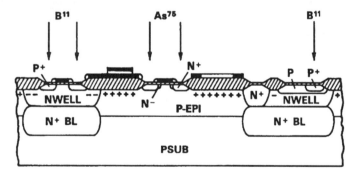

Figure 3.61 Definition of the CMOS source/drain and NPN base contact regions.

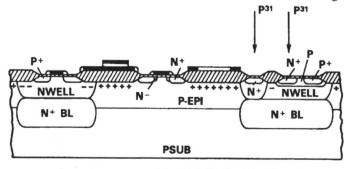

Figure 3.62 Formation of the n+ emitter and collector regions.

3.8.4.9 Source/Drain Formation

Using separate resist masking operations, the N+ and P+ source/drain regions are independently formed with arsenic and boron implants respectively. These implants are offset from the gate electrode by means of the sidewall spacer to reduce overlap capacitance (Fig. 3.61).

3.8.4.10 Emitter Formation

Using a resist mask, phosphorus is implanted to form the bipolar emitter and collector contact regions (Fig. 3.62).

3.8.4.11 Contact and Metal Formation

A doped oxide layer is deposited over the wafer, and contacts are patterned and dry-etched. A reflow operation can be done to improve metal step coverage over polysilicon steps and into contacts. A metal layer is then deposited, patterned, and etched. A protective overcoat is deposited and patterned to complete the process.

3.8.5 Future Process Issues in Analog BiCMOS

Since analog BiCMOS designs usually incorporate digital functions and are based on standard cell strategies, there will always be a push toward smaller feature sizes, particularly in the CMOS logic section, where digital standard cells are now moving into the sub-micron range. These digital processes utilize shallow heavily-doped twin-wells, very thin gate oxides, and silicides to improve device performance. However, as stated previously, thin oxides and high doping levels conflict with analog CMOS needs. Hence, separate wells and dual gate oxides may be required. Furthermore, the use of silicides will negate the use of polysilicon resistors unless additional process complexity is added. As can be seen, severe process trade-offs must be considered as analog BiCMOS processes are scaled to the micron and sub-micron level. Otherwise, prohibitively large mask counts will be required.

The use of deep diffusions such as N-well, N+ collector, and the N+ buried layer result in wasted silicon, due to excessive lateral diffusion. The use of deep trench isolation could significantly impact analog BiCMOS packing density. Again, the cost of adding this process module must be weighed against the impact in packing density and speed. As the push toward system chips continues, the need for integrating power components into existing high-performance analog BiCMOS processes will become more important. High voltage/current devices such as DMOS transistors may need to be comprehended. In summary, analog BiCMOS processes will continue to lag behind digital processes in terms of speed and packing density. However, this disparity will be reduced slightly as analog system power supply levels continue to fall.

Acknowledgements

The authors are indebted to several groups within Texas Instruments, Inc. for many of the achievements in BiCMOS technology that are highlighted in this chapter. Although individual contributors are too numerous to list here, the authors would like to particularly thank their colleagues in the Semiconductor Process and Design Center, Linear Circuits, New Product/Technology Department, Houston Bipolar Development, DLOGIC-1/DLOGIC-2 and T.I. Japan for their contributions to this work.

References

3.1] T. Ikeda, A. Wantanabe, Y. Nishio, I. Masuda, N. Tamba, M. Odaka and K. Ogiue, "High-speed BiCMOS technology with a buried twin-well structure," IEEE Trans. Electron devices, vol. ED-34, pp. 1304-1309, June 1987.

3.2] B. Bastani, C. Lage, L. Wong, J. Small, L. Bouknight and T. Bowman, "Advanced one micron BiCMOS technology for high speed 256K SRAM"s," Digest of Technical Papers, 1987 Symposium on VLSI Technology, pp. 41-42, May 1987.

3.3] R. Havemann, R. Eklund, R. Haken, D. Scott, H. Tran, P. Fung, T. Ham, D. Favreau and R. Virkus, "An 0.8 μm 256K BiCMOS SRAM technology," Digest of Technical Papers, 1987 International Electron devices Meeting, pp. 841-843, December 1987.

3.4] M-L. Chen, C-W. Leung, W. Cochran, R. Harney, A. Maury and H. Hey, "A high performance submicron CMOS process with self-aligned channel-stop and punchthrough implants," Digest of Technical Papers, 1986 International Electron Devices Meeting, pp. 256-259, December 1986.

3.5] R. Chapman, R. Haken, D. Bell, C. Wei, R. Havemann, T. Tang, T. Holloway and R. Gale, "An 0.8 μm CMOS technology for high performance logic applications," Digest of Technical Papers, 1987 International Electron Devices Meeting, pp. 362-365, December 1986.

3.6] J. Miyamoto, S. Saito, H. Momose, H. Shibata, K. Kanzaki and S. Kohyama, "A 1 μm n-well/bipolar technology for VLSI circuits," Digest of Technical Papers, 1983 International Electron Devices Meeting, pp. 63-66, December 1983.

3.7] H. Higuchi, G. Kitsukawa, T. Ikeda, Y. Nishio, N. Sasaki and K. Ogiue, "Performance and structures of scaled-down bipolar devices merged with CMOSFETS," Digest of Technical Papers, 1984 International Electron Devices Meeting, pp. 694-697, December 1984.

3.8] H. Tran, D. Scott, P. Fung, R. Havemann, R. Eklund, T. Ham, R. Haken and A. Shah, "An 8 ns 256K ECL SRAM with CMOS memory array and battery backup capability", IEEE J. Solid State Devices, vol. SC-23, pp. 1041-1047, October 1988.

3.9] R. de Werdt, P. van Attekum, H. den Blanken, L. de Bruin, F. op den Buijsch, A. Burgmans, T. Doan, H. Godon, M. Grief, W. Jansen, A. Jonkers, F. Klaassen, M. Pitt, P. van der Plas, A. Stolmeijer, R. Verhaar and J. Weaver, "A 1M SRAM with full CMOS cells fabricated in a 0.7 μm technology", Digest of Technical Papers, 1987 International Electron Devices Meeting, pp. 532-535, December 1987.

3.10] H-J. Bohm, L. Bernewitz, W. Bohm and R. Kopl, "Megaelectronvolt phosphorus implantation for bipolar devices," IEEE Trans. Electron Devices, vol. ED-35, no. 10, pp. 1616-1619, October 1988.

3.11] J. Borland, M. Gangani, R. Wise, S. Fong, Y. Oka and Y. Matsumoto, "Silicon epitaxial growth for advanced device structures," Solid State Technology, pp. 111-119, January 1988.

3.12] R. Herring, "Advances in reduced pressure silicon epitaxy," Solid State Technology, pp. 75-80, November 1979.

3.13] M. Graef and B. Leunissen, "Antimony, arsenic, phosphorus, and boron autodoping in silicon epitaxy," J. Electrochem. Soc., vol. 132, no. 8, pp. 1942-1954, August 1985.

3.14] G. Srinivasan, "Kinetics of lateral autodoping in silicon epitaxy," J. Electrochem. Soc., vol. 125, no. 1, pp. 146-151, January 1978.

3.15] R. Rung, H. Momose and Y. Nagakubo, "Deep trench isolated CMOS devices," Digest of Technical Papers, 1982 International Electron Devices Meeting, pp. 237-240, December 1982.

3.16] C. Teng, C. Slawinski and W. Hunter, "Defect generation in trench isolation," Digest of Technical Papers, 1984 International Electron Devices Meeting, pp. 586- 589, December 1984.

3.17] S. Suyama, T. Yachi and T. Serikawa, "A new self-aligned well-isolation technique for CMOS devices," IEEE Trans. Electron Devices, vol. ED-33, pp. 1672-1677, November 1986.

3.18] N. Kasai, N. Endo, A. Ishitani and H. Kitajima, "1/4 µm CMOS isolation technique using selective epitaxy," IEEE Trans. Electron Devices, vol. ED-34, pp. 1331-1336, June 1987.

3.19] K. Cham, S. Chiang, D. Wenocur and R. Rung, "Characterization and modeling of the trench surface inversion problem for the trench isolated CMOS technology," Digest of Technical Papers, 1984 International Electron Devices Meeting, pp. 23-26, December 1983.

3.20] J. Brighton, D. Verret, T. Ten Eyck, M. Welch, R. McMann, M. Torreno, A. Appel, M. Keleher, "Scaling issues in the evolution of ExCL bipolar technology," Digest of Technical Papers, IEEE 1988 Bipolar Circuits & Technology Meeting, pp. 121-124, September 1988.

3.21] K. Chiu, J. Moll and J. Manoliu, "A bird"s beak free local oxidation technology feasible for VLSI circuits fabrication," IEEE Trans. Electron Devices, vol. ED-29, pp. 536-540, April 1982.

3.22] K. Wang, S. Saller, W. Hunter, P. Chatterjee and P. Yang, "Direct moat isolation for VLSI," IEEE Trans. Electron Devices, vol. ED-29, pp. 541-547, April 1982.

3.23] N. Matsukawa, H. Nozawa, J. Matsunaga and S. Kohyama, "Selective polysilicon oxidation technology for VLSI isolation," IEEE Trans. Electron Devices, vol. ED-29, pp. 561-573, April 1982.

3.24] A. Alvarez, P. Meller, D. Schucker, F. Ormerod, J. Teplik, B. Tien and D. Maracas, "Technology considerations in BI-CMOS integrated circuits," International Conference on Computer Design, pp. 159-163, 1985.

3.25] T. Ning, P. Cook, R. Denard, C. Osburn, S. Schuster and H. Yu, "1 µm MOSFET VLSI technology: Part-IV - Hot electron design constraints," IEEE Trans. Electron Devices, vol. ED-26, pp. 346-353, 1979.

3.26] S. Ogura, P. Tsang, W. Walker, D. Critchlow, and J. Shepard, "Design and characteristics of the lightly doped drain source (LDD) insulated gate field effect transistor," IEEE Trans. Electron devices, vol. ED-27, pp. 1359-1367, 1980.

3.27] H. Katto, K. Okuyama, S. Meguro, R. Nagai, and S. Ikeda, "Hot carrier degradation modes and optimization of LDD MOSFETS," Digest of Technical Papers, 1984 International Electron Devices Meeting, pp. 774-777, December 1984.

3.28] K. Balasubramanyam, M. Hargrove, H. Hanafi, M. Lin, and D. Hoyniak, "Characterization of As-P double diffused drain structure," Digest of Technical Papers, 1984 International Electron Devices Meeting, pp. 782-785, December 1984.

3.29] S. Meguro, S. Ikeda, K. Nagasawa, A. Koike, T. Yasui, Y. Sakai and T. Hayashida, "Hi-CMOS III technology," Digest of Technical Papers, 1984 International Electron Devices Meeting, pp. 59-62, December 1984.

3.30] H. Nakashiba, I. Ishida, K. Aomura, and T. Nakamura, "An advanced PSA techology for high-speed bipolar LSI," IEEE Trans. Electron Devices, vol. ED-27, pp. 1390-1394, August 1980.

3.31] Y. Kobayashi, M. Oohayashi, K. Asayama, T. Ikeda, R. Hori and K. Itoh, "Bipolar CMOS merged structure for high speed M bit DRAM," Digest of Technical Papers, 1986 International Electron Devices Meeting, pp. 802-804, December 1986.

3.32] M. Brassington, M. El-Diwany, P. Tuntasood and R. Razouk, "An advanced submicron BiCMOS technology for VLSI applications," Digest of Technical Papers, 1988 Symposium on VLSI technology, pp. 89-90, May 1988.

3.33] K. Rajkanan, T. Gheewala and J. Diedrick, "A high-performance BiCMOS technology with double-polysilicon self-aligned bipolar devices," IEEE Electron Device Letters, vol. EDL-8, pp. 509-511, November 1987.

3.34] T. Yuzuriha, T. Yamaguchi and J. Lee, "Submicron Bipolar-CMOS technology using 16 GHz fT double-polysilicon bipolar devices," Digest of Technical Papers, 1988 International Electron Devices Meeting, pp. 748-751, December, 1988.

3.35] G. Kitsukawa, R. Hori, Y. Kawajiri, T. Watanabe, T. Kawahara, K. Itoh, Y. Kobayashi, M. Oohayashi, K. Asayama, T. Ikeda and H. Kawamoto, "An Experimental 1-Mbit BiCMOS DRAM," IEEE J. Solid-State Circuits, vol. SC-22, pp. 657-662, October 1987.

3.36] E. Chor, P. Ashburn and A. Brunnschweiler, "Emitter resistance of arsenic- and phosphorus-doped polysilicon emitter transistors," IEEE Electron Device Letters, vol. EDL-6, pp. 516-518, October 1985.

3.37] P. Ashburn, D. Roulston and C. Selvakumar, "Comparison of experimental and computed results on arsenic- and phosphorus-doped polysilicon emitter bipolar transistors," IEEE Trans. Electron Devices, vol. ED-34, pp. 1346-1353, June 1987.

3.38] B. Landau, B. Bastani, D. Haueisen, R. Lahri, S. Joshi, and J. Small, "Poly emitter bipolar transistor optimization for an advanced BiCMOS technology," Digest of Technical Papers, 1988 IEEE Bipolar Circuits and Technology Meeting, pp. 117-119, September 1988.

3.39] H. Park, K. Boyer, C. Clawson, G. Eiden, A. Tang, T. Yamaguchi, and J. Sachitano, "High-speed polysilicon emitter-base bipolar transistor," IEEE Electron Device Letters, vol. EDL-7, pp. 658-660, December 1986.

3.40] E. Greeneich and K. McLaughlin, "Analysis and characterization of BiCMOS for high-speed digital logic," IEEE J. Solid-State Circuits, vol. 23, pp. 558-565, April 1988.

3.41] R. Chapman, D. Bell, R. Eklund, R. Havemann, M. Harward and R. Haken, "Submicron BiCMOS well design for optimum circuit performance," Digest of Technical Papers, 1988 International Electron Devices Meeting, pp. 756-759, December 1988.

3.42] T. Ikeda, A. Watanabe, Y. Nishio, I. Masuda, N. Tamba, M. Odaka and K. Ogiue, "High-speed BiCMOS technology with a buried twin well structure," IEEE Trans. Electron Devices, vol. ED-34, pp. 1304-1310, June 1987.

3.43] D. Tang and P. Solomon, "Bipolar transistor design for optimized power-delay logic circuits," IEEE Journal of Solid-State Circuits, vol. SC-14, pp. 679-684, August 1979.

3.44] S. Konaka, Y. Amemiya, K. Sakuma and T. Sakai, "A 20 ps/G Bipolar IC using advanced SST with collector ion implantation," Extended Abstracts of the 19th Conference on Solid State Devices and Materials, pp. 331-334, 1987.

3.45] D. Scott, W. Hunter and H. Shichijo, "A new transmission line model for silicided source/drain diffusions: Impact on VLSI circuits," IEEE Trans. Electron Devices, vol. ED-29, pp. 651-661, April 1982.

3.46] M. Alperin, T. Holloway, R. Haken, C. Gosmeyer, R. Karnaugh and W. Parmantie, "Development of the self-aligned titanium silicide process for VLSI applications," IEEE Trans. Electron devices, vol. ED-32, pp. 141-149, February 1985.

3.47] C. Lau, Y. See, D. Scott, J. Bridges, S. Perna and R. Davies, "Titanium disilicide self-aligned source/drain and gate technology," Digest of Technical Papers, 1982 International Electron Devices Meeting, pp. 774-777, December 1982.

3.48] R. Haken, "Application of the self-aligned titanium silicide process to VLSI NMOS and CMOS technologies", J. Vac. Sci. Technol. B 3(6), pp. 1657-1663, Nov/Dec. 1985.

3.49] D. Scott, R. Chapman, C. Wei, S. Mahant-Shetti, R. Haken and T. Holloway, "Titanium disilicide contact resistivity and its impact on 1 μm CMOS circuit performance," IEEE Trans. Electron Devices, vol. ED-34, pp. 562-574, March 1987.

3.50] T. Tang, C. Wei, R. Haken, T. Holloway, L. Hite and T. Blake, "Titanium nitride local interconnect technology for VLSI," IEEE Trans. Electron Devices, vol. ED-34, pp. 682-688, March 1987.

3.51] O. Kudoh, H. Ooka, I. Sakai, M. Saitoh, J. Ozaki and M. Kikuchi, "A new full CMOS SRAM cell structure," Digest of Technical Papers, 1984 International Electron Devices Meeting, pp. 67-70, December 1984.

3.52] C. Kaanta, W. Cote, J. Cronib, K. Holland, P. Lee and T. Wright, "Submicron wiring technology with tungsten and planarization," Digest of Technical Papers, 1987 International Electron Devices Meeting, pp. 67-70, December 1987.

3.53] T. Bonifield, R. Gale, B. Shen, G. Smith and C. Huffman, "A 1 micron design rule double level metallization process," Proceedings of the IEEE 1986 VLSI Multilevel Interconnection Conference, pp. 71-77, June 1986.

3.54] T. Bonifield, S. Crank, R. Gale, J. Graham, C. Huffman, B. Jucha, G. Smith, M. Yao, S. Aoyama, Y. Imamura, K. Hamamoto, H. Kawasaki, T. Kaeriyama, Y. Miyai, M. Nishimura and M. Utsugi, "A 2 micron pitch triple level metal process using CVD tungsten", Digest of Technical Papers, 1988 Symposium on VLSI Technology, pp. 101-102, May 1988.

3.55] T. Yuzuriha and S. Early, "Failure mechanisms in a 4 micron pitch gold IC metallization process," Proceedings of the IEEE 1986 VLSI Multilevel Interconnection Conference, pp. 146-152, June 1986.

3.56] C. Hu, S. Chang, M. Small and J. Lewis, "Diffusion barrier studies for copper," Proceedings of the IEEE 1986 VLSI Multilevel Interconnection Conference, pp. 181-184, June 1986.

3.57] A. Alvarez, P. Meller and B. Tien, "2 micron merged bipolar-CMOS technology," Digest of Technical Papers, 1984 International Electron Devices Meeting, pp. 761-764, December 1984.

3.58] M. Polinsky and S. Graf, "MOS-Bipolar monolithic integrated circuit technology," IEEE Trans. Electron Devices, vol. ED-20, pp. 239-244, March 1973.

3.59] M. Polinsky, O Schade Jr. and J. Keller, "CMOS-Bipolar monolithic integrated-circuit technology," Digest of Technical Papers, 1973 International Electron Devices Meeting, pp. 229-231, December 1973.

3.60] H. Lin, "Comparison of input offset voltage of differential amplifiers using bipolar transistors and field-effect transistors," IEEE J. Solid-State Circuits, vol. SC-5, pp. 126-129, June 1970.

3.61] H. De Man, R. Vanparys and R. Cuppens, "A low input capacitance voltage follower in a compatible silicon-gate MOS-Bipolar technology," IEEE J. Solid-State Circuits, vol. SC-12, pp. 217-223, June 1977.

3.62] "Si-gate CMOS programmable op amps rival bipolars," Electronic Products, p. 34, March 4, 1983.

3.63] E. Snow and B. Deal, "Polarization phenomena and other properties of phosphosilicate glass films on silicon," J. Electrochem. Society, vol. 113, no. 3, pp. 263-269, March 1966.

3.64] S. Miller, "Advances in process development lead to new architectures in data converters," Electro '83, pp. 1-5, April 1983.

3.65] T. Lindenfelser, D. Fertig, M. Schmidt and K. Perttula, "A 12-volt analog/digital BiCMOS process,"IEEE Proceedings of Bipolar Circuits and Technology Meeting, pp. 184-187, 1987.

3.66] K. Sato, K. Shimizu, Y. Nakamura, K. Oka, F. Nakamura and T. Kimura, "A novel Bi-CMOS technology with upward and downward diffusion technique," Electronics and Communication in Japan, Part 2, vol. 69, no. 7, pp. 1-8, 1986.

3.67] S. Weber, "TI soups up LinCMOS process with 20-V bipolar transistors," Electronics, Feb. 4, 1988, pp. 59-60.

3.68] T. Guy and D. Grant, "Complete DAC chip weds true monotonicity to 16-bit resolution," Electronic Design, June 14, 1984, pp. 273-282.

3.69] P. Gray and R. Meyer, Analysis and Design of Analog Integrated Circuits, pp. 116-117, New York: Wiley 1984.

3.70] M. Nanba, T. Shiba, T. Nakamura and T. Toyabe, "An analytical and experimental investigation of the cutoff frequency Ft of high-dpeed bipolar transistors," IEEE Trans. Electron Devices, vol. ED-35, pp. 1021-1028, July 1988.

3.71] R. Troutman, "VLSI limitiations from drain-induced barrier lowering," IEEE Trans. Electron Devices, vol. ED-27, pp. 461-468, April 1979.

3.72] P. Gray and R. Meyer, Analysis and Design of Analog Integrated Circuits, pp. 389-394 and pp. 705-9, New York: Wiley, 1984.

3.73] H. Katto, Y. Kamigaki and Y. Itoh, "MOSFET's with reduced low frequency 1/f noise," Proc. 6th Conference on Solid State Devices, pp. 243-248, Tokyo, 1974.

3.74] A. Grove, Physics and Technology of Semiconductor Devices, pp. 230-234, New York: Wiley, 1967.

3.75] C. Anagnostopoulos, P. Zeitzoff, K. Wong and B. Brandt, An isolated vertical npn bipolar transistor in an n-well CMOS process, Digest of Technical Papers, 1984 International Electron Devices Meeting, pp. 588-593, December 1984.

Chapter 4

Process Reliability

R. Lahri, S.P. Joshi, and B. Bastani
(National Semiconductor Corporation)

4.1 Introduction

The reliability of BiCMOS process is a key issue in establishing its viability. Some concerns have been raised regarding the increased process complexity and the compatibility of fabricating CMOS and bipolar devices on the same chip. Performance and process tradeoffs are expected as these two device types are merged with minimal additional process steps. Therefore, it is imperative that device reliability issues are integrated with the process architecture and device design issues.

The increased complexity of BiCMOS process also provides it with certain unique features which considerably enhance its reliability. These features include the presence of a thin epitaxial layer, self aligned twin wells, twin buried layers and the availability of a wider variety of device structures. These process features also determine the device performance. Therefore, optimizing process architecture will minimize performance and reliability tradeoffs. This methodology, also called reliability by design, is being increasingly used to build reliability into VLSI processes.

In this chapter, the reliability by design approach has been utilized to address reliability concerns in advanced BiCMOS processes. The discussion focuses on soft errors, hot carrier effects in CMOS and bipolar devices, gate oxide integrity, electromigration, latchup and ESD. As is demonstrated in this chapter, an optimized BiCMOS process can achieve reliability which is superior to purely CMOS or bipolar processes. New techniques to evaluate process reliability like Wafer Level Reliability (WLR) testing and Process Control Monitor (PCM) stressing are also described.

4.2 Reliability by Design

Reliability by design is a relatively new concept in which reliability is built into the process rather than evaluated after the process has been developed. It is an integrated approach aimed at proactively identifying and solving process reliability issues with minimal performance tradeoffs. It addresses device, process, circuit and packaging related reliability issues at the test chip level so that potential reliability concerns are identified and solved early in the process development cycle. This methodology includes optimizing the process architecture, defining appropriate design rules, laying out and testing various reliability intensive test structures and setting up wafer level reliability monitors for quick feedback. Process margins are established for various reliability issues like MOSFET hot carrier degradation vs gate length, bipolar device stability vs emitter current density, alpha particle induced charge collection vs supply voltage etc. Also, process optimization is carried out to minimize potential reliability concerns like plasma damage to gate oxide, mobile ion contamination, film stresses in the backend etc. Device and subcircuit level stresses like burn-in are carried out to establish the reliability of the process prior to an actual product qualification. This type of stressing, also called PCM (Process Control Monitor) stressing, has the advantage of easy diagnostics at the test chip level and provides valuable information regarding process marginalities.

4.2.1 Reliability and Performance Tradeoffs

In any VLSI process, optimization of device parameters is driven by circuit performance and reliability requirements. Often these requirements are conflicting, which results in tradeoffs. However, by proper process architecture, such tradeoffs can be minimized. In a BiCMOS process, such optimizations can be difficult since CMOS and bipolar devices have differing requirements. For example as discussed in chapter II, thin (1.0 - 1.5 μm) low resistivity (0.3 - 0.5 ohm-cm) epitaxial layers are necessary for high performance switching NPN transistors. Whereas, CMOS devices prefer thick (≈ 10 μm), high resistivity (10 - 15 ohm-cm) epitaxial layers for low source drain junction capacitance and low body effect. Another example is the requirement of low temperature backend (planarization and interconnect formation) processing for CMOS devices to keep source/drain junctions shallow. On the other hand, in the case of poly emitter bipolar transistors, if the backend temperature is too low (< 900°C), the oxide beneath the polysilicon forming the emitter may not breakdown, causing variable and undesirably high emitter resistance. It follows that the BiCMOS process optimization must be done judiciously. The number of process parameters that can be decoupled are limited by the need to maintain process simplicity.

Process frontend (prior to planarization and interconnect formation) has the greatest impact on device performance and reliability. The epitaxial layer com-

bined with heavily doped N + buried layer is used in bipolar technology to minimize NPN collector resistance and improve breakdown voltage. As an added
benefit, the epitaxial/N + buried layer also increases the alpha of lateral PNP
transistors. DC and AC NPN performance depends critically on the choice of
epitaxial layer doping and thickness. For CMOS devices, thinner epi increases
the junction capacitance and degrades its performance though significant improvement in the latchup immunity are seen due to reduced N-well resistance
[4.1]. Fig. 4.1 shows the gate delay for CMOS and bipolar ring oscillators with
varying epitaxial layer thickness [4.2]. As expected, thinner epi improves
bipolar performance whereas CMOS performance is degraded due to increased source/drain junction capacitance. Therefore, an optimization of epi
/N + buried layer thickness will include a trade off between the performance of
CMOS and bipolar devices as well as the latchup immunity of CMOS devices.

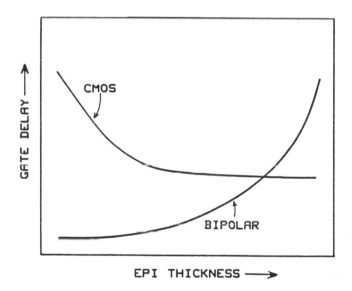

Figure 4.1: Gate delay for CMOS and Bipolar ring oscillators with varying
epitaxial layer thickness. From [4.2].

P + buried layer dose is another important front end process parameter. P +
buried layer is primarily used to decrease NPN island spacing by increasing the
N-well to N-well punchthrough. This buried layer also increases soft error immunity by reducing alpha particle induced charge collection. However, increased P + buried layer dose degrades the performance of N-channel FETs
due to increased source/drain junction capacitance and increased body factor.
The effect on CMOS ring oscillator performance is more dominant for thinner
epitaxial layers as shown in Fig. 4.2 [4.2]. Once again a trade off is expected in
order to optimize P + buried layer dose.

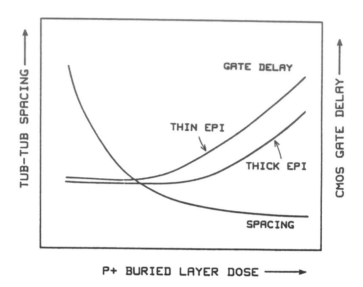

Figure 4.2: Effect of P + buried layer dose on the well to well spacing and CMOS gate delay. From [4.2]. © IEEE.

Reducing MOSFET gate lengths improves their current drive capability and increases CMOS ring oscillator speed as demonstrated by Fig. 4.3. However, narrow gate length devices also become sensitive to short channel effects like V_t dependence on gate length, drain induced barrier lowering, channel hot electron induced degradation etc. Another important effect occurs in memory circuits where the node capacitance decreases with gate width as gate oxide capacitance associated with that node is a significant part of the total node capacitance. This has the effect of increasing soft error rate. Thus, performance and reliability issues are very interrelated and should be addressed through an integrated approach.

4.2.2 Monitoring Reliability

Traditionally process qualification has involved subjecting a product circuit to burn-in, temperature cycling, temperature humidity testing etc. These tests are designed to accelerate failure modes that are likely to occur during normal device operation. Activation energies for various failure mechanisms are determined to predict circuit lifetime under normal operating conditions. Table 4.1 provides activation energy for typical IC failure mechanisms.

Whereas this approach provides an excellent indication of process reliability, it has the drawback of finding reliability concerns after the process has been developed and characterized. The solutions become more complex to implement towards the final stages of process development and often result in compromises between performance and reliability. Also, once a process is qualified based on this approach, it is difficult to maintain the same processing conditions later on without utilizing some kind of reliability monitors. Therefore, new approaches to monitoring reliability are not only important to assess process

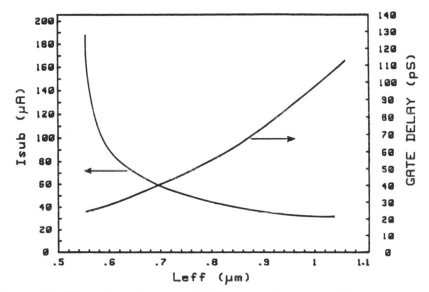

Figure 4.3: Effect of gate length on CMOS ring oscillator gate delay and N-channel MOSFET substrate current

Table 4.1: Activation Energy For Typical IC Failure Mechanisms

Failure Mechanisms	Activation Energy E_a (eV)
Ionic Contamination	1.0
Dielectric Breakdown	0.3
Hot Carrier Trapping in Oxides	- 0.06
Electromigration (Al-Cu-Si)	0.7
Gold-Aluminum Intermetallic Growth	1.0

during the development phase for quick feedback but also to maintain the process within a set of control limits once it has been qualified. Two approaches are getting increasing recognition in this area. First is the wafer level reliability (WLR) which looks for reliability concerns at the wafer level and provides quick feedback in case processing conditions drift out of the control limits. The second approach is to carry out the stressing of process control monitors (PCMs) to look for device instabilities early in the process development cycle. These two techniques have become an integral part of the reliability by design concept.

Wafer Level Reliability. Wafer level reliability tests focus on oxide integrity, mobile ion contamination, electromigration of interconnects and contacts/via, dielectric TDDB (Time Dependent Dielectric Breakdown) etc. WLR tests involve very high acceleration factors since the tests need to be completed in a reasonably short time ($<$ 1 min.) so as not to affect wafer throughput at parametric testing and provide quick feedback. Failures which normally occur in tens of years are induced in a matter of few minutes through accelerated stressing. Therefore, it is very important that these tests are designed such that they reflect actual failure modes and that no new failure modes are introduced as a result of such highly accelerated stressing. Conventional electromigration, oxide TDDB, and hot electron tests apply relatively small stress levels and hence take longer to complete. These conventional burn-in tests induce the same failure modes as seen under normal operating conditions. Activation energies for various failure modes are established and product lifetime are calculated using closed form expressions for acceleration factors. The temperature acceleration factor (A_t) for a failure mode, for example, is given by the Arrhenius equation:

$$A_t = Exp\,[\,E_a/k\,(1/T_s - 1/T_o)] \qquad (4.1)$$

where T_s is the stress temperature, T_o is the device operating temperature, E_a is the activation energy for the failure mode, and k is the Boltzmann constant.

However, in the case of WLR testing, prediction of product lifetime is more complex as the acceleration factors involved are extremely high. WLR tests are correlated with conventional reliability tests to establish baselines and control limits. This baseline provides a powerful tool in monitoring reliability on a routine basis and providing quick feedback for process control. Field failure data is often used to focus on major reliability concerns. These tests are also used to identify and optimize process steps that affect gate oxide integrity, metal step coverage, hot electron susceptibility of MOSFETs etc [4.3]. WLR tests related to gate oxide testing and electromigration are described in the following sections of this chapter.

Process Control Monitor Stressing. Process Control Monitor (PCM) stressing is carried out to get an indication of process reliability early in the process development cycle. Device parametric shifts, mobile ion contamination and threshold instabilities are identified by subjecting individual transistors, small transistor arrays and circuits like ring oscillators and delay strings to burn-in, temperature cycling, temperature-humidity environment etc. A major advantage of PCM stressing is the ease of failure diagnostics since failure mode identification and characterization is much simpler at the device level. PCM stressing is also receiving a lot of attention in the ASIC business where the cost of separately qualifying a series of very similar products can be very high. In such cases, PCM stressing in combination with product qualification data on one product circuit can be used to qualify other similar circuits. Once PCM stressing has shown the basic process reliability at the device and subcircuit level, further product qualification may not be necessary for each option.

PCM stressing is conceptually simple but great care needs to be taken in defining device stress conditions. If proper biasing schemes are not used, the devices may be subjected to extreme stresses and may reflect instabilities which are not expected in an actual circuit environment. For example, bipolar transistors have been reported to show beta instability for very high emitter currents [4.4] but in an actual circuit environment these current levels may be unrealistic. Normally PCM stressing is carried out under DC conditions but circuits like ring oscillators can provide a good indication of device stability under AC conditions.

4.3 Built-in Immunity To Soft Errors

Alpha particle induced soft errors in memories are becoming a major reliability concern with higher levels of system integration and reducing device dimensions in VLSI technology. When an alpha particle strikes an internal node of a memory cell, it can cause that particular cell to loose its charge state. The source of alpha particles is predominantly package material (both ceramic and plastic) which typically has emission flux of 0.1 - 1.0 alpha/cm^2*hr. Even though this flux is very low, soft errors are still detected in a system comprising of a large number (> 1000) of such memory chips. System designers compensate for these random errors with various error detection schemes at the expense of system performance. The most desired error correction scheme is parity check but soft error rate (SER) needs to be fairly low (less than 100 FITs) in order to utilize this scheme without significantly affecting the system performance. If the SER is very low, the system designer may choose to eliminate error correction schemes altogether to boost performance. Various approaches are being utilized to reduce SER in VLSI circuits. For dynamic memories (DRAMs), SER can be reduced by increasing the amount of stored charge (30-50 fC) that defines the memory state. Novel techniques like trench capacitor cells and high-C cells have been proposed to increase storage node capacitance [4.5, 4.6]. The static memory (SRAM) cell, however, does not have a capacitive element. The storage node capacitance, in case of a SRAM cell, is determined by parasitic capacitances which decrease as cell size is reduced. In such cases, either the radiation source flux is minimized by applying die coat [4.7] or structural protection barriers are utilized [4.8] to reduce charge collection efficiency. The P + buried layer in combination with the thin epitaxial layer in a BiCMOS process provides an excellent structural barrier to alpha particle induced charge collection and significantly reduces SER in BiCMOS circuits.

4.3.1 Alpha Particle Induced Charge Collection

When an alpha particle strikes silicon, it travels and loses energy along a straight path. The length of travel depends on the energy of the alpha particle. The energy lost by the incident particle goes to create high energy electron-hole (e-h) pairs. These hot e-h pairs decay into plasmons which in turn create more e-h pairs. This complex thermalization process occurs within a time of the order of picoseconds and finally results in one e-h pair for every 3.6 eV energy lost by an alpha particle [4.9]. These e-h pairs form a cylindrical column of radius 0.1 μm with a net neutral charge. If an alpha particle penetrates the depletion region of a p-n junction, the column of carriers disturbs the depletion region. Due to the electric field in the region, generated electrons drift towards more positive potential and holes towards more negative potential. At the same time, e-h pairs in the quasi-neutral substrate region start to separate by one Debye length after one dielectric relaxation time. The net charge becomes the slow moving holes located in the center and fast moving electrons located at the edge of the free carrier column. The equipotential lines of the depletion region rapidly spread downward and envelope the entire length of alpha track. Fig. 4.4 shows the potential distortion due to the alpha strike. This phenomenon is called the field funneling effect.

The field funneling causes carriers in the track to be collected rapidly and efficiently by drift rather than slowly by diffusion, where most of the generated carriers recombine. Once the charge collection process has completed, the disturbed field relaxes back to the original position as the junction re-establishes itself. Field funneling not only enhances the charge collection but also increases other reliability problems like induced latchup, source-drain shorts [4.10, 4.11], emitter collector shorts,and leakage between trench capacitors.

The field funneling phenomenon is characterized by a parameter called the effective funneling length (L_f) which is an indicator of the extent of depletion region distortion as a result of an alpha strike on a p-n junction. Higher funneling length leads to more efficient charge collection, resulting in higher soft error rates. As an example, the funneling effect is equivalent to the total collection of all charges for a distance of about 10 μm beyond the original depletion layer for a device on a 14 ohm-cm substrate which is biased at 15 V and struck by a 4.3 MeV alpha particle [4.12]. A three dimensional device simulator CADDETH [4.13] has been used to calculate L_f [4.14]. L_f depends on the substrate concentration (N_A), size of the p-n junction, and the incidence angle of alpha particle. Fig. 4.5 shows the dependence of L_f on substrate impurity concentration with p-n junction size as a parameter. L_f decreases as N_A increases.

This dependence is reduced as the junction size becomes larger. Smaller junction size reduces the funneling length since a depletion layer with a smaller area is better able to recover from the potential distortion. For the non-vertical alpha incidence condition, the e-h pairs generated within the depletion

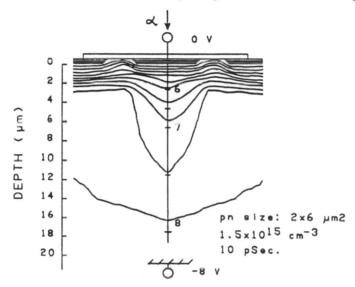

Figure 4.4: The potential distribution due to the funneling effect. From [4.9]. © IEEE.

region do not move along the alpha particle track, but along the electric field within the depletion region. This provides a rather small potential distortion near the side edge of the depletion region and results in much smaller L_f.

Proximity of adjacent cells in a memory array also affects the charge collection process. Funneling induced potential distortion extends to adjacent cells where some of the generated carriers are shared, thereby reducing the effective funneling length. Charge sharing between the latch and the pass transistors in a four transistor (4-T) memory cell is discussed in Chapter 6.

4.3.2 Role of Epitaxial Layer and Buried Layer

Traditionally, P-well has been used as a structural barrier to reduce charge collection in CMOS VLSI circuits. However, a P-well alone has not proven to be sufficient as funneling fields can still inject the generated carriers into the storage node. Charge collection up to 130 fC has been reported [4.8] even with fairly high (2.5E16 cm^{-3}) P-well surface concentration. However, if the P-well

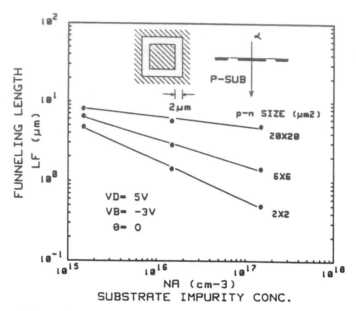

Figure 4.5: Funneling length vs. P-N junction size relationship. From [4.9].
© IEEE.

resides in n-type substrate, it provides a more effective barrier. This is because
the depletion region associated with the reverse biased P-well/substrate junction reduces the alpha particle induced charge collection. A P-type buried layer
below the storage node, with peak concentration in the 1E17 - 1E18 cm^{-3} range,
has been reported [4.15] to reduce charge collection by a factor of five. Soft
error rate depends exponentially on the amount of collected charge. Thus a
reduction in charge collection by a factor of five reduces SER by three orders
of magnitude. In the case of a high concentration P + buried layer, the funneling field, which is a strong function of substrate doping, collapses at the buried
layer with very minimal penetration into it. The charge collection occurs only
from carriers generated within the depletion region of the P-N junction and
those generated in the region between the junction and the buried layer.

BiCMOS process architecture has the inherent advantage of thin epitaxial
layer and a built-in P + buried layer. Charge collection as low as 25 fC has been
obtained without any extra processing steps. Fig. 4.6 shows the charge collection as a function of buried layer dose.

Once the peak charge collection is reduced below the critical charge necessary to flip the cell, the circuit becomes virtually immune to soft errors. Increased buried layer dose and reduced epitaxial layer thickness, however,
increase the source drain junction capacitance [4.16] and, therefore, a compromise becomes necessary.

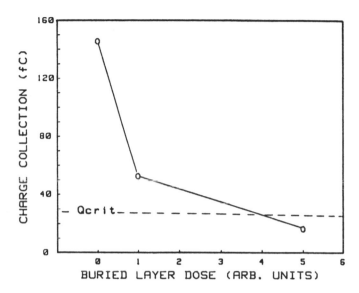

Figure 4.6: Maximum N+ diode charge collection for varying buried layer dose.

4.3.3 Soft Errors in BiCMOS SRAMs

As mentioned earlier, the storage node capacitance for a SRAM cell is determined by the parasitic capacitance associated with that node. This parasitic node capacitance is a function of the source/drain diffusion capacitance, gate width of the latch transistor, junction sidewall capacitance etc. The gate width of the latch transistor is a critical parameter in determining the node storage capacitance and hence the SER sensitivity of a 4-T SRAM cell. Fig. 4.7 shows the effect of n-channel gate width on the soft error rate of a 256K SRAM using 4-T cell.

Other factors important in determining SER of a SRAM cell include the supply voltage (V_{EE}) levels, threshold mismatch of the latch transistors, load resistors and the cell usage statistics. As V_{EE} is reduced, SRAM cell noise margin as well as the storage node charge is reduced. Both these factors contribute to higher SER as shown in Fig. 4.7. Cell stability and SER sensitivity are interlinked since a decreased noise margin adversely affects soft error immunity. A graphical technique utilizing transient circuit simulation of SRAM cells has been developed [4.17] to investigate the margin of these cells to alpha particle induced disturbance. This technique also facilitates comparisons between different cell designs and different assumptions about usage statistics to find optimum SRAM cell from the stability and SER considerations.

4.3.4 Other Protection Techniques

As the size of SRAM cell is further reduced for denser and faster memories, parasitic node capacitance will decrease to an extent that P+ buried layer may not be able to provide sufficient protection against alpha particle induced

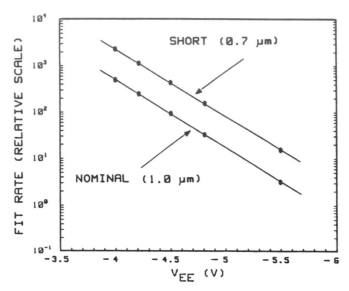

Figure 4.7: Effect of N-channel gate width on SER of a 256K BiCMOS SRAM.

charge collection. A new soft error immune BS (Buried Source) memory cell has been proposed [4.18] for mega-bit SRAMs. The BS cell, as shown in Fig. 4.8, increases the storage node capacitance and decreases the charge collection by changing the structure of the driver MOSFET from a lateral type to a vertical one with a buried source region in the substrate. Since the gate electrode of the driver MOSFET of the BS cell is formed in the trench, the parasitic capacitance is larger than that of the conventional cell. N+ buried layer serves as the source region. The charges generated in the substrate by alpha particles are collected in the source region. A factor of two reduction in the amount of collected charge has been reported with the BS cell. However, process complexity is a drawback of this approach.

A more conventional approach to reducing SER is to overcoat memory chips with a contamination free material which will absorb the incident alpha particles. Polyimide materials have been extensively used as die coats for this purpose. Typically, a 2.5 - 3.0 mil thick coating has been found to be sufficient to stop alpha particles with energies normally encountered in ceramic packages (up to 8.78 MeV). The excellent heat resistance characteristics of these

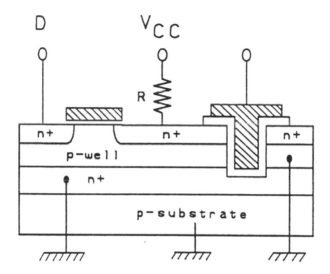

Figure 4.8: Schematic cross-section of a buried source SRAM cell. From
[4.18]. © IEEE.

materials have been reported which can be selectively coated only over the sensitive parts of the device, leaving the scribe lines and pad areas open for dicing
and wirebonding respectively. In addition, polyimides provide a strong
mechanical protection for the die during subsequent handling.

4.4 Gate Oxide Integrity and Hot Electron Degradation in MOSFETs

The two most important reliability concerns for MOSFETs are gate oxide
integrity and hot electron degradation. Poor quality oxides can affect circuit
functionality, increase leakage or shift device parameters. A common concern
in a BiCMOS process is that the MOS transistors are typically formed prior to
bipolar transistors and subsequent processing may adversely affect gate oxide
integrity. Hot electron degradation of N-channel MOSFETs normally increases their threshold voltage (V_t), and reduces transconductance (G_m) and
current drive capability. These parametric shifts in MOSFETs affect circuit
performance and, in extreme cases, circuit functionality. Hot electron degrada-

tion is a strong function of device stress conditions and therefore it is important to simulate the actual circuit environment for device stressing in order to accurately predict circuit lifetime due to this mechanism. In this section, both gate oxide integrity and hot electron degradation issues are discussed.

4.4.1 Gate Oxide Integrity

The intrinsic breakdown of gate dielectrics under high field conditions as well as low field defect related breakdown has been the subject of a great deal of experimental work [4.19, 4.20, 4.21]. Time dependent dielectric breakdown (TDDB) is considered to be one of the major failure modes for MOS ICs. Conceptually, TDDB process can be divided into two stages. During the first build-up stage, localized high field/ high current density regions are formed as a result of charge trapping within the gate oxide. Charge trapping occurs at localized spots and the trapped carriers are reported to be mostly holes [4.22]. When the local current density or field reaches a critical value, a rapid runaway stage begins which eventually results in breakdown. The runaway stage takes place in a very short period of time ($< 1\mu s$). Therefore, the time necessary to reach the runaway stage determines the lifetime of the oxides. Time for charge build-up is greatly reduced when defects or ionic contamination are present. For example, Na+ contamination can greatly accelerate the breakdown process. Particulates or silicon defects are known to be sources of localized high fields which weaken the gate oxide. Radiation damage to gate oxides during plasma processing is also reported to reduce breakdown voltage of MOS gate oxides [4.23].

The activation energy for TDDB comprises of two terms: temperature dependent and electric field dependent. The temperature acceleration factor (A_t) is given by the Arrhenius equation (eq. 4.1). The activation energy for this mechanism is approximately 0.03 eV. It is clear that temperature alone is not a good parameter to accelerate oxide related fails during burn-in. TDDB is a stronger function of electric field across the oxide. Hu et. al. have reported [4.24] that the electric field acceleration factor (A_{ef}) for TDDB is given by the following equation:

$$A_{ef} = Exp\ [\beta(1/E_s - 1/E_0)] \tag{4.2}$$

where E_s is the stress field and E_o is the desired operating field (both in MV/cm). The exponent is found to be 100-135 cm/MV. These activation energy numbers are dependent on the thickness of gate oxide and its quality. The thermal and electric field acceleration factors can be combined so that TDDB failures can be projected for desired operating conditions from the accelerated stress data. Table 4.2 summarizes the thermal and field acceleration factors for a 256K SRAM based on ambient temperatures. For more accurate calculations, junction temperatures should be considered instead.

Another important parameter to characterize oxide breakdown due to charge injection is the "charge-to-breakdown" or Q_{bd}. Q_{bd} is determined by forcing a constant current through the oxide and noting the time for breakdown. It is given by the following expression:

$$Q_{bd} = \int_0^{t_{bd}} J \, dt \qquad (4.3)$$

where J is the current density and t_{bd} is the time necessary for breakdown i.e. the time to complete the buildup stage. Q_{bd} has been found to be an excellent indicator of gate oxide quality and is reported to correlate with charge trapping during hot electron injection [4.25]. Haas et al have reported Q_{bd} values in the range 40-60 Coulomb/cm^2 for a BiCMOS process. High Q_{bd} values in combination with low gate oxide defect density numbers show that the gate oxide quality in a BiCMOS process is not affected by subsequent processing to make bipolar transistors [4.3].

TABLE 4.2: Thermal and Field Acceleration Factors for a BiCMOS SRAM

	Ambient Temp.	Field
Stress conditions:	150°C	3.125 MV/cm
Operating conditions:	85°C	2.75 MV/cm

Temperature acceleration factor (A_t):	4.5
Field acceleration factor (A_{ef}):	78.5
Overall acceleration factor:	353.25

4.4.2 DC Hot Electron Characteristics

As MOSFET channel lengths are reduced for higher current drive capability, gate oxide thickness and junction depth are decreased while channel doping levels are increased to maintain a reasonable threshold voltage. The supply voltage, however, is not decreased to maintain compatibility with system voltage requirements. Therefore, very high electric fields (~0.5-0.7 MV/cm) are generated near the drain edge of the MOSFET. Such intense electric fields cause hot carriers due to impact ionization within the channel pinchoff region. Generated holes move towards the substrate and give rise to substrate current (I_{sub}). Substrate current (I_{sub}) is a function of gate and drain voltage and an indicator of electric field at the drain edge as given by the following equation [4.26]:

$$I_{sub} = I_d \, \alpha \, \Delta L \qquad\qquad (4.4)$$

where I_d is the drain current and ΔL is the length of the pinch-off region. α is the ionization coefficient, the number of electron-hole pairs generated per unit distance and is a strong function of the lateral electric field at the drain edge. At sufficiently high I_{sub} levels, the source junction can get forward biased and induce latchup. Substrate current induced latchup is discussed in section 4.7. Some of the generated electrons are injected into the gate oxide under the influence of positive gate bias. Most of the injected carriers are accelerated through the gate oxide and are measured as gate current. However, a small percentage of these carriers get trapped in the gate oxide and cause device performance degradation. This effect has been traditionally called hot electron degradation, even though both holes and electrons have been postulated to contribute to the degradation mechanism. P-channel transistors are less susceptible to hot electron degradation than N-channel transistors because P+ source drain junctions tend to be deeper (reducing electric field at the drain edge) and the channel current tends to be lower due to lower mobility of holes. Also, the Si-SiO$_2$ barrier to hole injection is much higher. Therefore most of the work is directed towards the study of hot electron degradation in N-channel MOSFETs.

Hot carrier degradation in a N-channel MOSFET characteristics is manifested by an increase in threshold voltage, and a decrease in transconductance and current drive capability. The hot carrier effects can be minimized by (i) reducing the electric field near the drain junction (by using lightly doped drain (LDD) junctions), thus decreasing the emission probability of hot carriers and (ii) improving the gate oxide quality in terms of the density of hot carrier traps to reduce the effect of carriers injected in the gate oxide on device performance. Improving the quality of gate oxide must comprehend the optimization of both the gate oxidation process as well as subsequent processing.

DC hot electron degradation of a MOSFET is measured by stressing it at sufficiently high drain voltage and setting gate voltage to achieve maximum substrate current. The device lifetime criteria under DC stress conditions is fairly arbitrary and is typically chosen to be 10% degradation in G_m. In an actual circuit environment, this criteria will depend on the specific device biasing configurations as discussed in a section 4.4.4. By plotting device Gm degradation with time at various DC stress levels, time for 10% G_m degradation ($T_{10\%}$) is obtained as a function of substrate current. Fig. 4.9 shows DC hot electron degradation of an N-channel MOSFET (Leff = 0.7 μm) in a BiCMOS process.

DC device lifetime is calculated by extrapolating the $T_{10\%}$ numbers to substrate current levels at worst case operating conditions as shown in Fig. 4.10.

It must be emphasized here that $T_{10\%}$ is used as a figure of merit for comparing various device designs and may not be representative of device lifetime in an actual circuit operation.

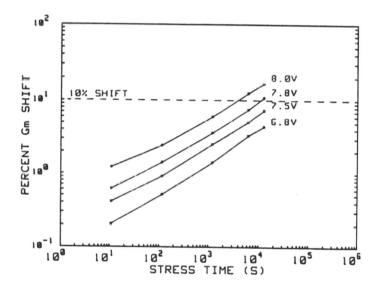

Figure 4.9: DC Hot electron degradation of an N-channel MOSFET (Leff = 0.7 μm) for varying drain bias stress.

4.4.3 Role of Backend Processing

Whereas electric field at the drain junction governs the hot carrier generation and injection, trapping of these carriers is determined by the gate oxide-silicon interface properties and the quality of the gate oxide itself. Subsequent processing to form interconnects can degrade Si-SiO2 interface properties. Plasma induced damage during RIE patterning and ashing operations are known to degrade oxide quality. The presence of hydrogen in plasma environment, normally seen during PECVD deposition of silicon nitride films for passivation is a major contributor to hot electron degradation susceptibility of MOSFETs. As shown in Fig. 4.11, degradation rate of MOSFETs is increased by two orders of magnitude by adding a silicon nitride passivation layer. Hydrogen in silicon nitride films resides in N-H bonds and Si-H bonds. Si-H bonds are weaker and break during subsequent high temperature (400-450°C) processing.

At these temperatures, hydrogen reacts with strained bonds at the Si-SiO$_2$ interface and creates an excess of Si-H or Si-OH bonds [4.27]. These bonds are precursors for bulk and interface trap generation processes. Recently some low hydrogen silicon nitride films have been developed which have low concentration of Si-H bonds. Application of these films for passivation has been reported to significantly improve hot electron degradation susceptibility of N-channel devices [4.28].

4.4.4 AC Hot Electron Degradation Characteristics

Though hot carrier degradation under DC stress conditions has been extensively used to characterize the susceptibility of N-channel devices to channel hot carriers, it is not a good indicator of the device performance in an actual circuit environment. Device biasing conditions, particularly those related to

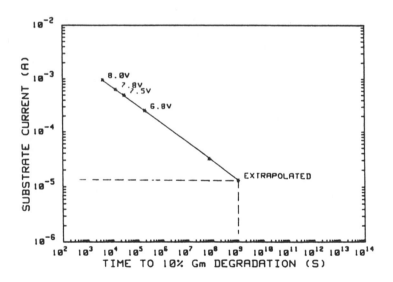

Figure 4.10: DC device lifetime extrapolation from hot carrier stressing data.

gate and drain determine the periods of maximum substrate current and the extent of hole and electron injection into the gate oxide. Therefore, it is imperative to carry out a detailed analysis of device biasing under an actual circuit operation in order to correlate device degradation with circuit performance degradation due to hot carriers. For example, pass transistor configurations in a SRAM circuit tend to show much higher device degradation than an inverter configuration. In the former case gate voltage is pulsed and drain voltage is held high, whereas for the latter both gate and drain voltages are pulsed. In order

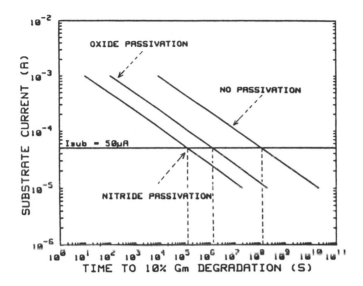

Figure 4.11: Effect of silicon nitride passivation on hot carrier degradation susceptibility of an N-channel MOSFET.

to predict device lifetime in an actual circuit environment, substrate current levels are used as a prime indicator of hot carrier degradation and AC conditions are simulated from the corresponding DC substrate current values [4.29]. However, transient substrate currents for very high slew rates (< 5 ns) may not be extrapolated from the DC values [4.30]. Because of fast pulse rise/fall times, transient substrate currents may not be able to reach their equilibrium DC values as shown in Fig. 4.12. Therefore, their impact during the transition period will be reduced.

In fact, for very high slew rates the displacement current component of the transient substrate current starts dominating the impact ionization current component (Fig. 4.13). This effect is specially important during the falling edge where the direction of overall substrate current favors hole injection. Hole injection followed by electron injection has been reported to cause severe hot carrier degradation [4.31].

Even though, 10% G_m degradation is considered to be a good figure of merit to optimize N-channel MOSFET device design, a realistic criteria to measure device degradation during hot electron stressing is dependent on the specific circuit configuration. For example, in an SRAM cell, the important device parameter for a pass transistor is Idsat whereas for the latch transistor, it is the linear region transconductance. Therefore, while estimating the susceptibility of a SRAM cell to hot carrier degradation, it is important to measure relevant device parameters in each biasing configuration. A generalized device lifetime criterion may result in under or overestimation of circuit susceptibility to hot carrier degradation.

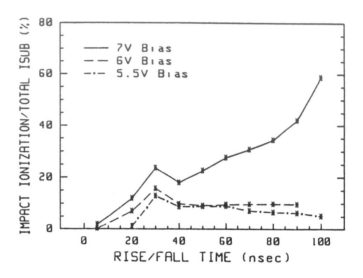

Figure 4.12: Variation in impact ionization current with rise/fall time.

4.5 Hot Carrier Effects in Bipolar Devices

When bipolar transistors are scaled-down, shallow junctions and high doping density in the emitter and base regions become imperative to maximize device performance. However, high electric fields at the emitter-base (e-b) junction decrease junction avalanche voltage and generate hot carriers due to impact ionization. This phenomenon is similar to hot carrier generation near the drain junction in MOSFETs. These high energy carriers are injected into the oxide region above the e-b junction under the influence of emitter bias and cause instabilities in device parametrics. It has become an important reliability concern in high performance poly emitter bipolar transistors. Such instabilities have been reported to exist under forward bias high injection region [4.4] as well as under reverse bias conditions [4.32]. In this section, hot carrier effects in scaled down poly emitter bipolar devices and their implications on circuit performance are discussed.

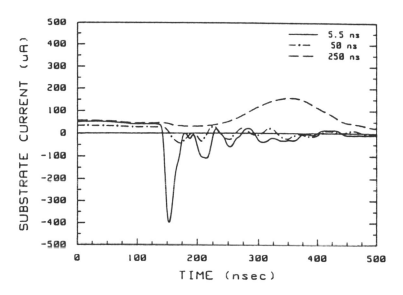

Figure 4.13: Transient substrate current variation during the falling edge for various slew rates.

4.5.1 Device Instability Due to High Injection Forward Bias

In a poly emitter bipolar transistor, the quality of the overlying oxide layer above the e-b junction (Fig. 4.14) determines the space charge recombination in the depletion region near the surface.

Fixed positive charges or donor type interface traps in the oxide deplete the surface of the underlying P-type base region and cause localized spreading of the depletion region. A wider depletion region increases space charge recombination leading to higher base current and lower beta. If the e-b junction is forward biased in the high injection mode, some of the injected electrons will neutralize the interface traps and reduce space charge recombination. This results in lower base current and higher beta. This beta instability has been observed during high temperature forward biasing of the poly emitter bipolar transistors (Fig. 4.15). These instabilities can be reduced by improving the quality of overlying oxide or by limiting forward emitter current($< 300 \ \mu A/\mu m^2$) [4.33].

4.5.2 Device Instability Due to Reverse Biasing of E-B Junction

In ECL logic, the emitter-base junction is generally not reverse biased. However, in a BiCMOS circuit where CMOS gates and bipolar drivers are interfaced, reverse biasing of the e-b junction can occur. Figure 4.16 shows a typical BiCMOS inverter (inset) and the HSPICE simulation waveform for the pullup transistor Q_1. Though the e-b junction is reversed biased only transiently, the stress on this transistor can be significant at high operating frequencies (100 MHz and above). Such stress conditions degrade the current gain (beta) of the device and can adversely affect circuit performance.

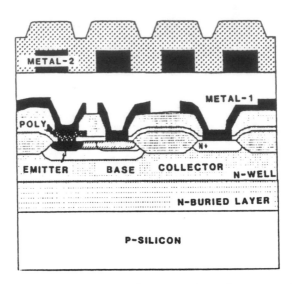

Figure 4.14: Schematic cross-section of a poly emitter bipolar transistor in the BiCMOS process.

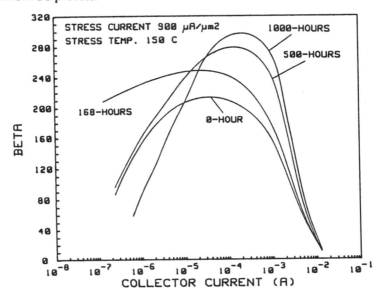

Figure 4.15: Beta variation under high current forward bias (@150°C) for a 2 μm X 4 μm poly emitter NPN transistor [4.33].

Constant voltage stressing of a poly emitter bipolar transistor indicates that, for a fixed stress duration, beta degradation increases with increasing reverse bias across the e-b junction [4.32]. With increasing reverse voltage levels, an increasingly large amount of current flows through the e-b junction. This reverse

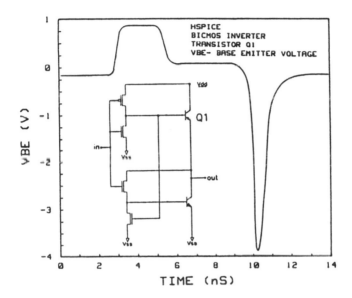

Figure 4.16: HSPICE simulation waveform for the pullup transistor (Q_1) of a BiCMOS inverter (inset).

current which is primarily confined to emitter perimeter region under the influence of localized electric field can generate hot carriers. These hot carriers damage the Si-SiO$_2$ interface at the junction and may get injected into the overlying oxide and generate a fixed oxide charge density and/or oxide trap density. This results in increased recombination in the space charge region and increases base current levels. Since the base-collector junction remains unaffected, forward beta degrades. The beta of a stressed device can be partially recovered by annealing in forming gas at 200°C or subjecting the device to a forward current soak at room temperature.

4.5.3 Charge to Degradation Model

Constant current stressing of reverse bias e-b junction reveals that the beta degradation is dependent on both the stress current and stress duration i.e. the total stress charge. Figure 4.17 shows an excellent correlation between stress charge and beta degradation for over six orders of stress current magnitude.

Beta degradation for a given stress charge (Q_R) in a simplified form can be expressed as :

$$d\beta/\beta = K_1 \, Log(Q_R) + K_2 \; ; Q_R > Q_{crit} \qquad (4.5)$$

K_1, K_2 and Q_{crit} are constants for a particular device design and are determined experimentally. The curve in Fig. 4.17 can be divided into two regions. In the initial stages of stressing (region I) interface traps play a major role in

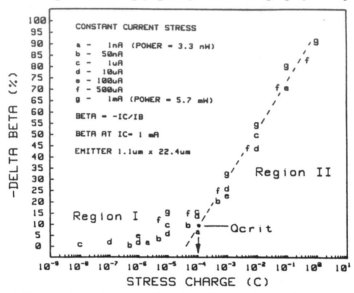

Figure 4.17: Beta degradation as a function of stress charge through a reverse bias e-b junction.

determining the extent of beta degradation. These donor type traps in conjunction with initial positive fixed oxide charges deplete the surface of underlying P-type base region. This increases recombination in the space charge region near the surface and results in higher base current and lower beta. The degradation in this region is soft as beta can be recovered by forward current soak or a high temperature (200°C) anneal. During the forward current soak, electrons occupy the donor type traps and make them neutral. Similarly, some of the positive fixed charge in the immediate vicinity of Si-SiO$_2$ interface can also be neutralized/compensated by low energy electrons leading to beta recovery. Beyond a certain critical charge (Q_{crit}) , a nonsaturating type of degradation can be sustained only if more and more traps are generated over a wider surface region, thus widening the corresponding space charge region.

Thus, charge to degradation model proposes a linear relationship between beta degradation and total charge through the reverse biased e-b junction above a certain critical charge. This relationship is very dependent on device architec-

ture and the field present at the e-b junction. Significantly lower beta degradation has been reported for Arsenic doped poly emitter transistors as compared
to Phosphorus doped poly emitter transistors [4.34].

4.5.4 AC vs DC Stressing

As shown in Fig. 4.18, the degradation under DC and pulsed DC conditions
is equivalent, for a given stress charge. Under AC conditions, however, beta
degradation is considerably reduced indicating that the forward current soak
reduces the positive charge build up near the e-b perimeter and beta integrity
is preserved longer. The reverse stress charge under typical AC conditions
remains below Q_{crit}. Therefore, in an actual circuit operation beta degradation
may be significantly lower than that obtained under DC or pulsed DC conditions. By simulating total reverse stress charge during one cycle of operation,
device degradation in a circuit environment can be predicted [4.32].

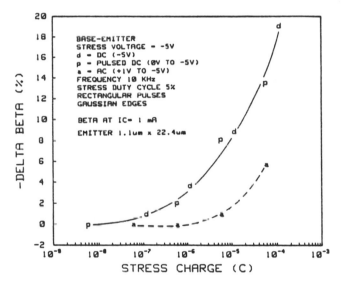

Figure 4.18: Dependence of beta degradation on reverse stress charge for a
poly emitter transistor.

4.6 Electromigration

Electromigration of interconnect lines has become an important reliability
concern in advanced VLSI circuits. Whereas device dimensions and interconnect linewidths are being reduced for higher density, the chip size and complexity is increasing to put more functions on a single chip. The higher

complexity and larger chip size require closely spaced long interconnect lines. As a result, the RC time delay, the IR voltage drop, power consumption and crosstalk noise associated with the interconnect lines becomes appreciable. Therefore an optimum interconnect for VLSI circuit applications would necessarily have improved electromigration resistance, low electrical resistivity, and silicon compatibility. Traditional Al-Si interconnects suffer from poor electromigration resistance, incompatibility with shallow junctions, poor step coverage, intermetal shorts due to hillock formation and photomasking problems due to the high surface reflectivity. Recently some multilayer interconnect systems have been reported [4.35, 4.36] which offer significant improvements in these areas. The multilayer interconnect systems, in combination with improved planarization in the process backend, have proven to be extremely reliable in advanced VLSI processes.

4.6.1 Multilayer Interconnect Systems

Addition of copper (0.5%) in Al-Si films has been reported to significantly improve electromigration resistance of these films [4.37]. Copper precipitates along the grain boundaries and suppresses aluminum migration. A small percentage of Cu (0.5%) has been found to be sufficient to increase the electromigration activation energy from 0.5 eV to 0.7 eV. A thin layer of titanium is often used on the top of Al-Cu-Si to reduce the surface reflectivity and suppress hillock formation. Reduced reflectivity eliminates reflective notching observed during patterning of aluminum lines. Hillock supression allows multiple levels of interconnects to be used without any concern about intermetal shorts. Presence of titanium also improves electromigration resistance as Ti films are compressive in nature and inhibit aluminum migration. However, the presence of titanium as an overlying layer mandates the use of silicon rich aluminum, since titanium has strong affinity towards silicon. Silicon content in the interconnect may need to be as high as 1-1.5%. For very shallow junctions (< 0.25 μm) even a higher concentration of silicon is not sufficient to avoid junction spiking. In such cases the use of a barrier beneath the Al-Cu-Si layer becomes necessary to avoid silicon migration from the junction to the overlying titanium layer. TiW has proven to be an effective barrier to silicon migration [4.38]. As discussed in the next section, such multilayer interconnect systems also exhibit significantly improved electromigration resistance.

4.6.2 Electromigration in Multilayer Interconnect Systems

In conventional Al-Si interconnect systems, aluminum migration through the grain boundaries causes open circuit fails resulting in relatively low median time to fail (10-15 years). As shown in Fig. 4.19, median time to fail improves significantly with the addition of a TiW barrier layer and titanium overlayer.

Another important feature of these multilayer interconnect systems is the virtual absence of open circuit fails. The resistance of interconnect lines may, however, increase by 20-30%. In order to understand this behavior, a re-examination of the basic electromigration phenomenon is necessary. In a pure Al-Si interconnect system, as aluminum migration starts, silicon accumulation in the direction of current flow is accompanied by void formation. The current density increases near the voids and enhances aluminum migration in localized areas. This eventually results in an open circuit failure. In a multilayer interconnect system, as voids start forming due to aluminum migration, the bulk of the current is carried by the underlying TiW layer in that region. TiW films are

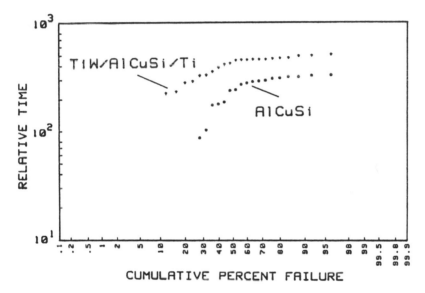

Figure 4.19: Lognormal plots showing superior electromigration resistance of Ti/ Al-Cu-Si/ TiW interconnects.

not susceptible to electromigration but are highly resistive. This causes localized heating and may anneal out the voids in the aluminum resulting in reduced resistance. This phenomenon may continue without the occurence of an open circuit failure. As shown in Fig. 4.20, the formation and subsequent annealing of the voids, during the stressing of interconnect lines to induce electromigration, results in resistance spiking. Behavior of Al-Cu-Si films is also shown for comparison.

Even though open circuit failures are unlikely in multilayer interconnect systems, they suffer from a higher incidence of intrametal or intermetal (in case of more than one level of interconnect) shorts. This happens as lateral extrusions of aluminum at the TiW/Al-Cu-Si interface occur due to mismatch in the mechanical stresses between the two films. Lateral extrusion is a manifestation of electromigration under the influence of lateral stresses. The activation energy for formation of lateral extrusions has been found to be similar (0.7 - 0.8 eV) to that associated with aluminum migration in the interconnect lines.

4.6.3 Role of Passivation Films

Passivation layer plays an important role in determining the electromigration resistance of the interconnect lines. Passivation films like silicon nitride or silicon oxynitride are necessarily deposited compressive in order to avoid cracks that are seen in tensile films. The compressive stress in the passivation films suppresses aluminum migration and significantly improves the electromigration resistance of the interconnect lines. However, if the compressive stress in-

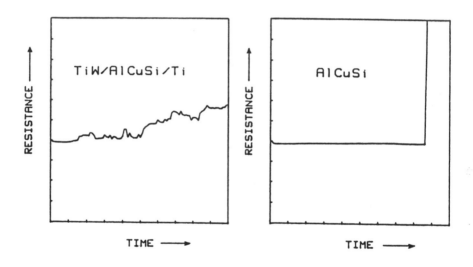

Figure 4.20: Resistance variation in Ti/Al-Cu-Si/TiW and Al-Cu-Si interconnect lines during the electromigration stressing.

creases above 3-4E9 dynes/cm^2, the underlying interconnect lines develop microvoids [4.39] which significantly reduce their electromigration resistance. In order to avoid the highly compressive stress of silicon nitride films, a phosphorus doped oxide layer is often used as a buffer beneath the nitride film.

4.7 Latchup in BiCMOS Circuits

One of the major problems encountered in implementing the use of CMOS

ICs has been their susceptibility to latchup [4.40, 4.41, 4.42]. Due to regenerative feedback of two parasitic bipolar transistors, a CMOS structure can be triggered and latched into a high-current state resulting in circuit nonfunctionality. BiCMOS process architecture offers some advantages towards reducing latchup susceptibility of the traditional CMOS structures. However, certain merged bipolar and CMOS structures in a BiCMOS circuit tend to be more susceptible to latchup and require special considerations. In this section, the latchup phenomenon is reviewed and its implications in BiCMOS technology are discussed.

4.7.1 Latchup Phenomenon

Latchup in standard CMOS (or CMOS part of BiCMOS) circuits results from the presence of parasitic NPNP paths, also called SCR paths. Figure 4.21 shows the lateral NPN and vertical PNP parasitic bipolar transistors. Normally these parasitic transistors are biased off. However, lateral current flow in the substrate (I_{RS}) and N-well (I_{RW}) can establish potential differences which turn on these parasitic transistors and establish an SCR path.

Figure 4.21 shows the first order equivalent circuit of the CMOS inverter in Fig. 4.20, where R_s is the lateral substrate resistance and R_w the lateral N-well resistance. Lateral current (I_{pp}) can result from (i) photocurrents due to ionizing radiation, (ii) displacement currents (C.dv/dt) from voltage spikes across the N-well junction, (iii) avalanche multiplication currents from the N-well junction due to overvoltage, or (iv) forward biasing of normally reverse biased junctions due to overshoots or externally applied voltages.

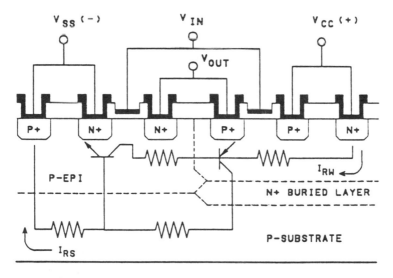

Figure 4.21: Schematic cross-section of a CMOS inverter showing the parasitic NPNP path for latchup.

A typical current-voltage latchup characteristic is shown in Figure 4.23. The holding current I_H is defined as the current at the holding voltage V_H. The trigger current (I_T) is defined as the current level where the SCR enters its negative resistance region.

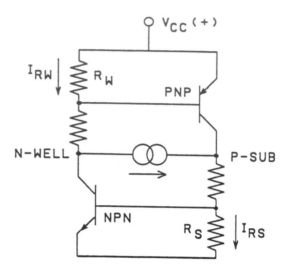

Figure 4.22: First order equivalent circuit of the CMOS inverter shown in Fig. 4.21. From [4.41]. © IEEE.

To sustain the latchup state, three conditions must be satisfied: (i) sufficient lateral voltage drop must occur to forward bias the e-b junction of both parasitic transistors, (ii) the transistor current gain product of the two parasitic transistors must exceed unity in order to achieve regeneration, and (iii) the bias supply must be capable of sourcing current greater than the holding current. If all these conditions are met, latchup occurs and the chip becomes nonfunctional due to the large current flow through it and the associated thermal runaway and electrical effects.

There are two reasons why BiCMOS structures are less susceptible to latchup. The first reason is the reduction of the substrate (I_{RS}) and N-well (I_{RW}) resistance. The presence of highly doped P+ and N+ buried layers significantly reduce these resistances. Especially significant is the N-well resistance which is reduced from the 1-2 Kohm/square range to below 50 ohm/square. Reducing R_s and R_w results in larger lateral currents being required for latchup initiation and higher holding currents. The second approach involves reducing the current gain (β) of the parasitic bipolars to increase latchup threshold. When the beta product ($\beta npn.\beta pnp$) falls below unity, latchup condition is impossible under all conditions. One way to reduce the

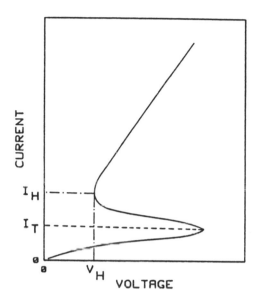

Figure 4.23: Typical current-voltage latchup characteristics.

beta product of the parasitic bipolars is to increase the N+ to P+ diffusion spacing [4.43]. An increase in N+ to P+ diffusion spacing is also shown to increase the holding voltage (Figure 4.24) which, in turn, improves the ability of the circuit to recover from a latchup [4.44].

Enhancing circuit latchup immunity by increasing the N-well to P+ spacing is not a preferred approach as it reduces the layout density. It is, however, commonly used in CMOS processes. In BiCMOS process, on the other hand, the presence of N+ buried layer significantly reduces beta of the vertical PNP by reducing carrier lifetime in the N-base region. Reduced beta and lower N-well resistance are two major reasons for significantly superior latchup immunity of BiCMOS process. For a 256K SRAM in 1.0μm BiCMOS technology, no latchup has been observed even when injected currents exceed +/- 500 mA.

4.7.2 Latchup in Merged Devices

Figure 4.25 shows an example of a merged CMOS/bipolar structure in BiCMOS technology [4.45]. At high collector current (I_c), the NPN collector debiases, saturating the transistor. This results in forward biasing the C-B junction and injecting holes into the substrate. These carriers form base current for parasitic NPN in the SCR latch and turn on the NPNP path. Fortunately the presence of N+ and P+ buried layers requires very large lateral current flow to initiate latchup. Using guard rings to sink minority carriers in the N-well and substrate further improves the latchup immunity of these structures.

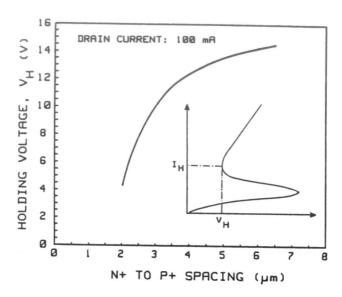

Figure 4.24: Holding voltage variation as a function of N+ to P+ spacing From [4.44]. © IEEE.

Figure 4.26 shows the latchup characteristics of another merged device where both NPN bipolar and PMOS transistors are in the same well. The schematic diagram is shown in the inset. Once the PMOSFET is turned ON, high voltage at the its source appears at the base of the NPN bipolar transistor and establishes a collector current (I_c) flow through it. This current flow continues even after the PMOSFET is turned off. The SCR action in this case is between the vertical NPN (which has a high beta) and lateral PNP which is a relatively good quality bipolar transistor. As soon as the PMOSFET is turned on, the internal P+ node (node A) is also elevated to the voltage at the external node. Since the base resistance is relatively high, the vertical NPN device is turned on and latchup is initiated. However, if the supply current is limited, latchup in these structures can be avoided.

In summary, BiCMOS process architecture offers superior latchup immunity for traditional CMOS structures. However, merged CMOS/bipolar structures should be carefully evaluated from latchup considerations and in some cases not allowed in the design rules.

4.8 ESD Protection

Electrostatic discharge (ESD) is a significant cause of device failure at all stages of IC testing, assembly, installation and field use. ESD can damage or

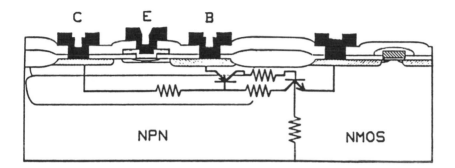

Figure 4.25: Latchup in a merged CMOS/Bipolar device. From [4.45].

destroy ICs which in turn can cause electronic system malfunction. CMOS devices tend to be more susceptible than bipolar devices to ESD. Circuit designers have used a wide variety of CMOS protection circuits [4.46, 4.47, 4.48]. BiCMOS technology offers a wider choice of devices for improved ESD protection. For measuring and characterizing ESD sensitivity, the industry looks to the military specification (MIL-STD-883, Method 3015.6) for standardization. However, this method only covers the "human body" model (HBM) which represents a situation where a human has obtained a static charge that is discharged into an IC when it is touched. Two other testing procedures are gaining increasing acceptance. First is the "charged device" (CD) model where a charged device is subsequently grounded, resulting in an ESD pulse through the device. Second is the "field induced" model which represents a situation where a charge is induced on it by an electrostatic field. It is very likely that an IC may see one or more of these situations during testing and assembly. An effective ESD test methodology will include all three types of simulators.

4.8.1 ESD Failure Models

Human Body Model. The HBM is the method most commonly specified in the United States for measuring and classifying the ICs for sensitivity to ESD. An equivalent circuit describing this model is shown in Fig. 4.27.

Using this model several investigators have determined thresholds for various device technologies [4.49, 4.50]. Typical protection capability of an IC under these conditions is in 2000 - 4000 volts range. With a charge of only 2000 volts, the human body stores approximately 0.2 millijoules of energy [4.51]. At discharge, this energy is dissipated in the body and the device resistance. With

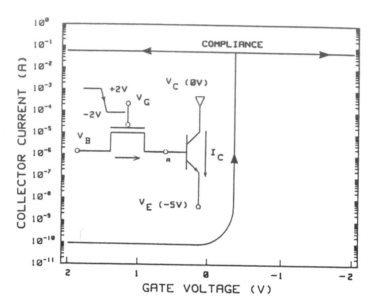

Figure 4.26: Latchup characteristics of a merged PMOS and NPN bipolar transistor structure (inset) in the same N-well.

the typical circuit elements shown in Fig. 4.27, this energy is released with time constants measured in tenths of microseconds, providing average power up to several kilowatts. Such short bursts of power in many cases are sufficient to melt small volumes of silicon and create small craters on the die surface. In some cases, ESD may only initiate a subtle weakness that gives rise to failure with subsequent device operation.

Charged Device Model. This model has gained large acceptance in Japan. It is associated with the device itself assuming charge on its lead frame or other conductive paths triboelectrically. Figure 4.27 shows the equivalent circuit for this model also.

With device capacitance in the 1 to 20 pF range, a device can store up to 100 microjoules of energy. With a low resistance path and lead frame inductance around 10 nH, this energy can be released on contact with ground giving rise to nanosecond discharge pulses with average power per pulse ranging from several hundred to several thousand watts. Such power levels are sufficient to degrade or destroy ICs.

Figure 4.28 compares the average power developed with human body model and charged body model for the same potential (2000 volts) across the device.

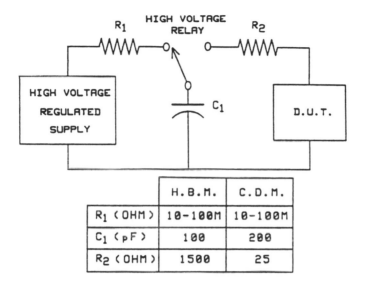

	H.B.M.	C.D.M.
R_1 (OHM)	10-100M	10-100M
C_1 (pF)	100	200
R_2 (OHM)	1500	25

Figure 4.27: Equivalent circuit describing the Human Body Model and the Charged Device Model. Relevant circuit parameters are summarized in the table (inset).

As is evident, average power available for damage during the charged device model is about 10 times greater than that for the human body transient.

Field Induced Model. This model is associated with a charged object. Electrostatic fields are associated with all charged objects and a potential gradient exists between the charged object and ground. IC package lead frames act as an antenna and induce potentials across MOS devices which can exceed the breakdown strength of the device dielectrics and cause device failure. Consider a device which is accidentally put in a charged plastic bin or placed in an area of high field environment. A charge displacement occurs on the conductive elements of the device, without causing any net charge. In many situations such as when assembling printed circuit boards, the device is picked up by hand. When a pin on the device is contacted, a discharge pulse occurs. This ESD is very much like that mentioned above for the charged device model except that the discharge results from an induced charge rather than a triboelectric charge. Failure analysis of devices that had a human body event show that the damage occurs at the pin contacted. However, failure and damage resulting from the charged device or induced charge event can occur internal to the circuitry or on pins remote from the one involved in the ESD.

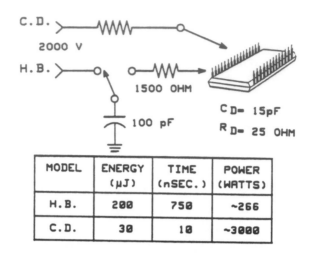

MODEL	ENERGY (µJ)	TIME (nSEC.)	POWER (WATTS)
H.B.	200	750	~266
C.D.	30	10	~3000

Figure 4.28: Comparison of the average power transferred to the device for the Human Body Model and Charge Device Model. The potential accross the device is same (2000 V) in both cases.

4.8.2 ESD Protection in BiCMOS Circuits

ESD protection circuits in BiCMOS technology are generally similar to those utilized in CMOS ICs. Figure 4.29 is an example of a typical input pad protection circuit. The parasitic resistance associated with the back to back diodes is also shown.

The diode series resistance under forward bias plays a critical role in determining the voltage developed across the MOS gate during an ESD zap. Presence of buried layers and availability of base-collector diodes in a BiCMOS process significantly reduces the series resistance and improves its ESD susceptibility. Thus, ESD susceptibility of BiCMOS devices can be superior to that of pure CMOS devices.

4.9 Summary

BiCMOS process architecture provides certain unique features which significantly enhance the process reliability. A new approach - reliability by design, which involves building reliability into the process through the optimization of its architecture, provides an excellent tool to address device reliability concerns at the process definition phase. The conventional approach of evaluating device reliability after the process has been developed suffers from the drawback that the solutions are difficult to implement and are often at the cost of performance.

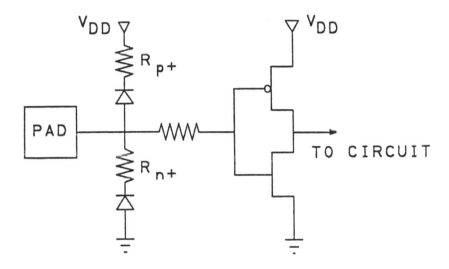

Figure 4.29: A typical ESD protection circuit in BiCMOS technology.

BiCMOS process provides significant improvements in alpha particle immunity, latchup and ESD susceptibility over the conventional CMOS process. Gate oxide integrity, in a process, where poly emitter bipolar transistors are formed after the MOS transistors, also does not apear to be a concern, if special processing techniques are implemented to avoid radiation induced damage to the oxide. Hot carrier effects in both CMOS and bipolar devices are minimized through device architecture optimization. Multilayer interconnect systems, utilizing refractory metals, provide excellent electromigration resistance in the state-of-the-art VLSI circuits.

A new set of reliability concerns will arise as the line widths are further reduced to enhance performance and density. The best approach to build reliability into future processes is to address the performance and reliability issues at the process definition phase. This will greatly minimize performance and reliability tradeoffs.

References — Process Reliability

4.1] T. Ikeda, T. Nagano, N. Momma, K. Miyata, H. Higuchi, M. Odaka, and K. Ogiue, "Advanced BiCMOS Technology for High Speed VLSI", IEDM Tech. Dig. pp. 408-411, 1986.

4.2] A. V. Alvarez, Presented at The Electronic Materials Symposium, 1987.

4.3] Fred Haas, Marshall Davis, Rajeeva Lahri, "Wafer Level Reliability Testing Using Keithley Parametric Test System", Wafer Level Reliability Workshop, Lake Tahoe, 1988.

4.4] T. C. Chen, C. Kaya, M. B. Ketchen, and T. H. Ning, "Reliability Analysis of Self-Aligned Bipolar Transistor Under Forward Active Current Stress" IEDM Tech. Dig. pp. 650-653, 1986.

4.5] G. A. Sai-Halasz et al., IEEE Trans. Electron Devices, vol. ED-29, no.4, pp. 725-731, April 1982.

4.6] M. Aoki et. al., Japan Journal of Applied Physics, vol. 21 supplement 21-1, pp. 73-78, 1982.

4.7] Kazumichi Mitsusada, Hisao Katto, Toru Toyabe, "Design for Alpha Immunity of MOS Dynamic RAMs", IEDM Tech. Dig., pp. 36-39, 1981.

4.8] Sai-Wai Fu, Amr M. Mohesen and Tim C. May, "Alpha-Particle-Induced Charge Collection Measurements and the Effectiveness of a Novel P-Well Protection Barrier on VLSI Memories", IEEE Transactions on Electron Devices, vol. ED-32, no. 1, Jan. 1985.

4.9] Chang-Ming Hsieh, Philip C. Murley and Redmond R. O'Brien, "Collection of Charge from Alpha-Particle Tracks in Silicon Devices", IEEE Transactions on Electron Devices, vol. ED-30, no. 6, June 1983.

4.10] Eiji Takeda, Dai Hisamoto, and Tohru Toyabe, "A new Soft Error Phenomenon in VLSIs" Proceedings International Reliability Physics Symposium, pp. 109-112, 1988.

4.11] D. Hisatomo, T. Toyabe, and E. Takeda, "Alpha-Particle-Induced Source-Drain Penetration Effects", Solid State Devices and Materials Conference, pp. 39-42, Tokyo, Aug. 1987.

4.12] C. M. Hsieh. P. C. Murley and R. R. O'Brien, "Dynamics of Charge Collection From Alpha Particle Tracks in Integrated Circuits", Proceedings International Reliability Physics Symposium, pp. 38-42, 1981.

4.13] Hiroo Masuda, Toru Toyabe, Hiroko Shukuri, K. Ohshima and Kiyoo Itoh, "A Full Three Dimensional Simulation on Alpha-Particle Induced DRAM Soft-Errors" IEDM Tech. Dig., pp. 496-499, 1985.

4.14] E. Takeda, K. Takeuchi, E. Yamasaki, T. Toyabe, K. Ohshima, and K. Itoh, "Effective Funneling Length in Alpha-Particle Induced Soft-Errors", Conference on Solid State Devices and Materials, pp. 311-314, Tokyo, 1986.

4.15] Hiroshi Momose, T. Wada, I. Kamohara, M. Isobe, J. Matasunga, and H. Nozawa, "A P-type Buried Layer for Protection Against Soft Errors in High Density CMOS Static RAMS" IEDM Tech. Dig., pp. 706-709, 1984.

4.16] Rajeeva Lahri, Craig Lage, Rick Jerome and Bami Bastani, "Inherent BiCMOS Soft Error Protection Through an Optimized P+ Buried Layer", First International BiCMOS Conference, Philadelphia May 1987.

4.17] B. Chappell, S. Schuster, G. Sai-Halasz "Stability and SER Analysis of Static RAM Cells" , IEEE Transactions on Electron Devices, vol. ED-32, no. 2, pp. 463-470, Feb. 1985.

4.18] M. Minami, Y. Wakui, H. Motsuki, and T. Nagano, "A New Soft Error Immune Static Memory Cell" IEDM Tech. Dig.,pp. 57-58, 1987.

4.19] Solomon "Breakdown in Silicon Oxide - A Review" J. Vac. Sci. Technology, vol. 14, no. 5 oct. 1977, pp 1122-1130.

4.20] I. C. Chen and Chenming Hu "Accelerated Testing of Time-Dependent Breakdown of SiO$_2$" IEEE Electron Device Letters, vol. EDL-8, no. 4, pp. 140-142, April 1987.

4.21] I. C. Chen, S. Holland, and C. Hu "Electrical Breakdown in Thin Gate and Tunneling Oxides" IEEE Transactions on Electron Devices, vol. ED-32, no. 2, pp. 413-422, Feb. 1985.

4.22] I. C. Chen, S. Holland, and C. Hu "Oxide Breakdown Dependence on Thickness and Hole Current-Enhanced Reliability of Ultra Thin Oxides" IEDM Tech. Dig., pp. 660-663, 1986.

4.23] T. Tsukura and S. Ueda "The Prevention Against Ashing Damage" Semiconductor World Japan, February 1987 (in Japanese).

4.24] C. Hu, "Thin Oxide Reliability", IEDM Tech. Dig. , pp. 368-371, 1985.

4.25] Marshall Davis and Rajeeva Lahri, IEEE Electron Device Letters, Vol. 9, No. 4, April 1988.

4.26] S. M. Sze "Physics of Semiconductor Devices", A Wiley-Interscience Publication, John Wiley & Sons, New York, p. 482

4.27] Yoav Nissan-Cohen, H. H. Woodbury, T. B. Gorczyca, C. Y. Wei "The Effect of Hydrogen on Hot Carrier Immunity, Radiation Hardness, and Gate Oxide Reliability in MOS Devices", Proceedings VLSI Tech. Symposium, pp. 37-38, 1988.

4.28] M. L. Chen, C. W. Leung, W. T. Cochran, S. Jain, H. P. W. Hey, H. Chew, and C. Dziuba "Hot Carrier Aging in Two Level Metal Processing" IEDM Tech. Dig., pp. 55-58, 1987.

4.29] T. Sakurai, M. Kakumu, and T. Iizuka "Hot Carrier Suppressed VLSI with Submicron Geometry" Proceedings International Solid State Circuits Conference, pp. 272-273, 1985.

4.30] Hai Wang, M. Davis, and R. Lahri "Transient Substrate Current Effects on N-Channel MOSFET Device lifetime" IEDM Tech. Dig., 1988 (to be presented)

4.31] Hai Wang, Steven Bibyk, and M. Davis, Presented at the ECS Fall Conference, 1988.

4.32] S. P. Joshi, R. Lahri, and C. Lage "Poly Emitter Bipolar Hot Carrier Effects in an Advanced BiCMOS Technology" IEDM Tech. Dig., pp. 182-185, 1987.

4.33] S. P. Joshi, National Semiconductor, Private Communication

4.34] B. Bastani, B. Landau, D. Hausien, R. Lahri, S. P. Joshi and Jim Small, Proceedings Bipolar Circuits and Technology Meeting, pp. 117-120, 1988.

4.35] D. Gardner, T. Michalka, K. Saraswat, T. Barbee, J. Mcvittie, and J. Meindl "Layered and Homogeneous Films Films of Aluminum and aluminum/Silicon with Titanium and Tungsten for Multilevel Interconnects" IEEE Journal of Solid State Circuits vol. SC-20, no. 1, pp. 94-103, Feb. 1985.

4.36] J. Maiz and B. Sabi "Electromigration Testing of Ti/Al-Si Metallization for Integrated Circuits" Proceedings International Reliability Physics Symposium, pp.145-153, 1987.

4.37] S. S. Iyer and C. Y. Ting "Electromigration Study of Al-Cu/Ti/Al-Cu System", Proceedings International Reliability Physics Symposium, pp. 273-278, 1984.

4.38] Fred Whitwer, Fred Haas, and Craig Lage "The Influence of Titanium Capped Aluminum on N + /P Junction Leakage" Proceedings IEEE VLSI Multilevel Interconnect Conference, pp. 484-490, 1988.

4.39] S. K. Groothius and W. H. Schroen, "Stress Related Failures Causing Open Metallization" Proceedings International Reliability Physics Symposium, pp. 1-8, 1987.

4.40] Ronald R. Troutman, "Recent Developments in CMOS Latchup" IEDM Tech. Dig., pp. 296-299, 1984.

4.41] D. B. Estreich, "The Physics and Modeling of latchup and CMOS Integrated Circuits" Ph.D. Dissertation, Stanford University, Stanford, CA October 1980.

4.42] R. S. Payne, W. N. Grant, and W. J. Bertram "Elimination of Latchup in Bulk CMOS", IEDM Tech. Dig., pp. 248-251, 1980.

4.43] A. R. Alvarez, J. Teplik, D. W. Schucker, T. Hulseweh, H. B. Liang, M. Dydyk, I. Rahim, "Second Generation BiCMOS Gate Array Technology" Proceedings IEEE Bipolar Circuits and Technology Meeting, pp. 113-117, 1987.

4.44] J. Manoliu, "Isolated Topics for High-Density BiCMOS", Proceedings of Bipolar and BiCMOS VLSI Technology Symposium, 1987.

4.45] C. Lage, National Semiconductor Corp., Private communication.

4.46] C. Duvvury, R. N. Rountree, and L. S. White, "A Summary of Most Effective Electrostatic Discharge protection Circuits for MOS Memories and their Observed Failure Modes", Proceedings International Reliability Physics Symposium, pp. 181-184.

4.47] T. V. Hulett, "On Chip Protection of NMOS Devices", Electrical Overstress/Electrostatic Discharge Symposium Proc. EOS-3, pp. 90-96, Sep. 1981.

4.48] C. Duvvury, R. N. Rountree, Y. Fong, R. A. McPhee, "ESD Phenomena and Protection Issues in CMOS Output Buffers",

4.49] D. C. Wunch, R. R. Bell, "Determination of Threshold Failure Levels of Semiconductor Diodes and Transistors Due to Pulse Voltages", IEEE Trans. Nucl. Sci. NS-15, 1968.

4.50] L. A. Schreier, "Electrostatic Damage Susceptibility of Semiconductor Devices", Proceedings International Reliability Physics Symposium, 1978.

4.51] B. A. Unger, "Electrostatic Discharge failures of Semiconductor Devices", Proceedings International Reliability Physics Symposium, pp. 193-199, 1981.

Chapter 5

Digital Design

K. Deierling (Dallas Semiconductor)

5.0 Introduction

The development of advanced BiCMOS processes has given the design engineer new flexibility with regards to basic digital design techniques. Conventional TTL, ECL, and CMOS technologies each have intrinsic strengths and weaknesses and thus each has established suitable applications in both standard "glue" logic as well as the newer application specific markets. BiCMOS technology allows fabrication of monolithic circuits which use all of the above design techniques as well as new hybrid BiCMOS circuits which take advantage of the strengths of the individual technologies.

The marriage of bipolar and CMOS technologies, though not always ideal, has provided a symbiotic design relationship yielding optimal circuit performance weighed against the expense of greater process complexity. Several advantages of CMOS circuits which have been the driving force in the technology's growth include the characteristic rail to rail output capability, extremely low quiescent power consumption, high density, and effectively infinite device input impedance; its major limitations are poor load driving capability; and performance degradation over voltage, process and temperature. Bipolar TTL design, on the other hand, has the advantage of better load driving capability and better performance over temperature and voltage at the expense of lower density and greater power consumption. Lastly emitter coupled logic (ECL), provides the fastest popular bipolar logic available, but also consumes the most power.

Each of these technologies' progression is potentially aided by the advent of BiCMOS.

5.1 CMOS vs. BiCMOS

Today's I.C. market is dominated by CMOS technology for several simple reasons: primary among these are the, generally acceptable speed performance, relatively straightforward processing, and the practically null static power consumption of CMOS circuits. BiCMOS circuits exhibit similar D.C. static characteristics to those of CMOS but enhance the A.C. drive characteristics at the expense of more complex processing and lower per gate density (a meaningful area comparison, however, must take into account the fact that BiCMOS circuitry is denser in terms of transconductance or load driving capability). It is relevant to review basic CMOS static and dynamic performance and compare to that of BiCMOS, so that appropriate use of BiCMOS circuitry is employed.

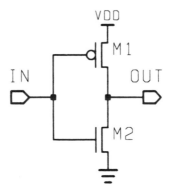

Figure 5.1 CMOS Inverter

5.1.1 Static Characteristics

The basic CMOS inverter (Fig. 5.1) has been well described elsewhere [1,2] and a basic knowledge of PMOS, NMOS, and bipolar transistors is assumed. The quiescent static characteristics of a representative BiCMOS inverter circuit and the simple CMOS implementation are compared below. A feeling for the almost zero

D.C. current drain of CMOS circuits can be gleaned from battery
backed memory products which incorporate low leakage (typically less
than 500nA) 256K SRAM memories (well over 1.5 million MOS
devices) and lithium batteries to insure data retention in the absence
of power for greater than ten years.

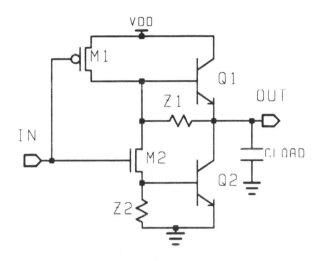

Figure 5.2 BiCMOS Inverter

The most common type of BiCMOS circuit uses bipolar
transistors to drive output loads, and CMOS to perform the logic
functions. The complimentary MOS devices provide for the extremely
low static current. The BiCMOS circuit of Fig. 5.2 is representative
of the general class of circuits currently available and rivals the static
characteristics of CMOS circuits. This circuit has a true VSS to VDD
output swing (this is not true of most BiCMOS circuits) with the
passive resistors Z1 and Z2 providing rail to rail swing once the
bipolar devices have been turned off.

Examining the BiCMOS circuit consider first a low at the input.
The PMOS device is then on and a highly conductive channel exists to
source base current into the pullup device Q1. The NMOS device, on
the other hand is cut off, therefore there is no D.C. path to supply
base current to Q2 so that resistor Z2 will turn off the common
emitter NPN device Q2. Since only A.C. current flows through the
output node the series path of MOS device M1 and resistor Z1 will

pull the output effectively to VDD. Thus the BiCMOS V_{OH} is identical to that of CMOS: typically 5.0V. An analogous situation exists with a high voltage provided to the input. In this case M2 is able to provide base current to Q2 which will discharge the load within a V_{BE} of ground. The series path through Z1, M2, and Z2 will discharge the load all the way to ground. In this case the PMOS device is cutoff thus Q1 has no D.C. base current and is in fact kept off by M2 and Z2 pulling the base voltage low. Thus the BiCMOS circuit's static characteristics rival those of the CMOS circuits, combining a noise margin of 5V (assuming only capacitive loading) and D.C. standby current confined to leakage currents presented by reversed biased pn junctions.

There have been several variations of the general circuit of Fig. 5.2 used in practice and each has different static and A.C. behaviour. Several common variations in use [3,4,5] are shown in Fig. 5.3.

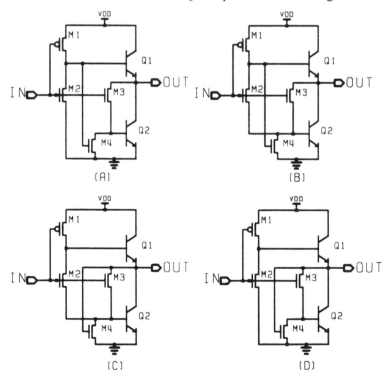

Figure 5.3 Several BiCMOS Inverter Implementations

It is important to note that the circuits in Fig. 5.3 do not have the full VSS to VDD static logic swing of CMOS circuits. For these circuits the V_{OL} is statically limited to a V_{BE} above ground (although A.C. effects cause actual circuits to pull down to the V_{CESAT} of the bipolar device typically in the $100 - 200$ mV range). The V_{OH} of these circuits is constrained by the V_{BEON} voltage of the bipolar device Q1 thereby limiting the output to VDD $- V_{BEON}$. In reality the bipolar device will continue to source current to the load with a logarithmic decrease as the output voltage increases and as a first order approximation the output voltage V_{OH} will rise to within a few hundred millivolts of the VDD supply. Considering increasingly shorter gate lengths and diminished threshold voltages, this is a serious limitation of these circuits which, without good process control, can lead to substantial subthreshold weak inversion static currents.

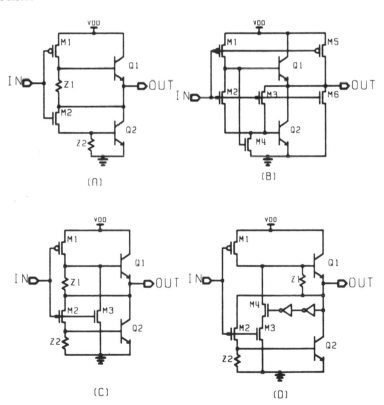

Figure 5.4 Full Rail BiCMOS Inverters

Additional circuits in use [3,6] are shown in Fig. 5.4 which do in fact achieve full rail swing although 5.4A passively and 5.4B only by the addition of the parallel CMOS inverter of M5 and M6. Even more complicated BiCMOS circuits have been proposed [7] which provide rail to rail swing and faster propagation delays. These circuits require more devices (Fig. 5.4C and D) and are variations to the circuits shown in Fig. 5.3. The circuit of Fig. 5.4B utilizes weak CMOS feedback inverters to increase propagation speeds. For a high to low input transition the NMOS device M4 is initially off so that M1 current is supplied to the base of Q1 without being shunted to ground, also improving power dissipation. For both circuits full rail swing is only passively achieved through resistors.

5.1.2 Dynamic Characteristics

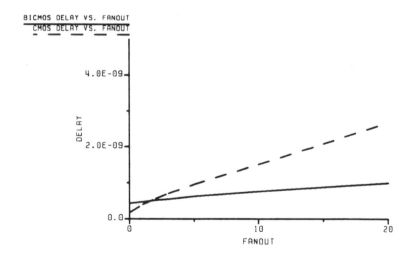

Figure 5.5 Typical Nand Delays vs. Fanout
BiCMOS NAND gate (Fig. 5.12) with Wp=30uM, Wn1=26uM, Wn2=20uM. CMOS NAND gate Wp=30uM, Wn=26uM. All lengths 1.5uM.

Next the dynamic performance of the simple CMOS inverter is compared to a BiCMOS inverter. The BiCMOS circuit performs better when driving high fanout loads as can be seen in a graph of typical propagation delay of both BiCMOS and CMOS NAND gates as a function of fanout (Fig. 5.5). Note that the crossover point beyond which the BiCMOS circuit is faster occurs at approximately a fanout of two demonstrating that even at relatively low fanouts the BiCMOS performance is superior to that of CMOS.

Figure 5.5 clearly points out the advantage of BiCMOS circuits when driving heavy capacitive loads and an analysis of the causes of this performance advantage are merited. When driving a heavy capacitive load the key factor determining switching speed is the current drive characteristics of the inverter. It is a well known fact that for equal area devices a bipolar device has a much greater transconductance than a MOS device [8]. A second important distinction between the MOS and bipolar current drive characteristics is the fact that the bipolar device is typically operated as a current controlled current source (with gain beta) while the MOS device operates as a voltage controlled current source. Consider a low to high transition at the input of the inverter such that as the input voltage rises above the threshold voltage of the NMOS device it turns on and begins to sink current thereby discharging the capacitive load. At this point the NMOS device is saturated since the drain potential is at 5V. As the input voltage rises and the drain voltage begins falling the NMOS device will enter the linear region, when V_{OUT} reaches a threshold below the gate (for a fast rising edge this occurs near V_{OUT} = V_{DD} − V_{TH} (⁻4.0V − ⁻4.5V). For the rest of the falling output transition the MOS device is in the linear region. The drain current of a MOS transistor in the linear region (neglecting second order effects) is expressed as:

$$I_d = \mu \bullet C_{ox} [2(V_{gs} - V_t)V_{ds} - V_{ds}^2]$$

The important factor to note is that as the output voltage falls (which is also the drain to source voltage of the NMOS transistor) the drain current decreases, approximately linearly at small drain voltages where the V_{ds}^2 term may be neglected. Contrast this to a bipolar transistor whose collector current characteristic is given simply:

$$I_c = I_s [e^{V_{be}/V_t} - 1]$$

where Vt is defined Vt=KT/Q and is typically 26mV at room temperature. Furthermore the collector current is linearly related to the supplied base current such that $I_c = \beta \ I_b$.

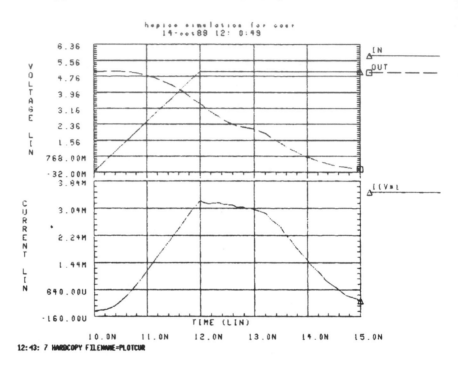

Figure 5.6 CMOS Discharge Current

CMOS inverter discharge current for Z_p =30uM, Z_n =25uM, L_p =L_p =1.5uM, fanout=10.

    To first order the collector current of common emitter connected bipolar transistor is independent of its collector voltage. In reality base width modulation causes the collector current to decrease slightly with decreasing collector voltage with a characteristic output impedance given by $R_o = V_a/I_c$ where V_a defines the Early Voltage (typically in the range 15V – 25V for modern BiCMOS NPN transistors). Thus the output impedance of a bipolar transistor conducting a large transient current of 5mA is still only 3–5kΩ which is insignificant for this analysis. The MOS transistor relevant characteristic impedance however is found as the change in $\Delta V_{ds}/\Delta I_d$, which is the much larger value:

$$R = \mu \bullet C_{ox} \bullet W/L[(V_{gs} - V_t) - V_{ds}]$$

approximately V_{ds}/I_d for small V_{ds}.

The rapid decrease of current with decreasing output voltage is readily apparent when the output current of a CMOS inverter (Fig. 5.6) is compared to the collector discharge current of a bipolar transistor from a BiCMOS inverter (Fig. 5.7) both driving high fanout loads.

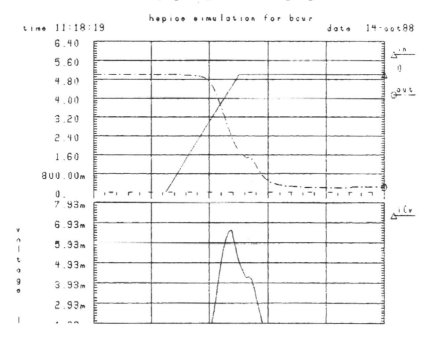

Figure 5.7 BiCMOS Discharge Current

BiCMOS inverter discharge current for A_E =24uM^2, Z_p =30uM, Z_n =25uM, L_p =L_p =1.5uM, fanout=10.

A similar effect favors BiCMOS circuits when considering worst case temperature, power supply voltages, and process variation [9]. Typically worst case delay specs must take into account voltages 10% below nominal VDD and junction temperatures of 100C or higher. The diminished gate to source voltage as well as drain to source voltage directly reduces switching current for CMOS circuits thus increasing delay times. Although in a BiCMOS circuit the MOS supplied base drive is reduced, since the bipolar device still remains saturated the switching current is much less markedly affected, therefore delays are less sensitive to diminished supply voltages. This is particularly important when considering full CMOS scaling since voltages must necessarily scale downward as a result of punchthrough and hot electron considerations. Similarly CMOS delays increase more than corresponding BiCMOS delays with temperature increases since phonon scattering causes a reduction in carrier mobility thereby directly affecting MOS drain current. The bipolar diffusion current is less affected than the drift current of the MOS devices. Although for the BiCMOS circuit, the bipolar transistor's base drive is decreased, this results in a much smaller effect on the delay time (Fig. 5.8).

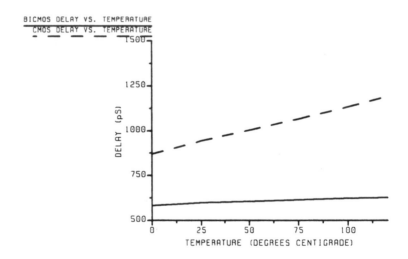

Figure 5.8 BiCMOS and CMOS Delays vs. Temperature
BiCMOS and CMOS NAND gate (fanout=5) typical propagation delays. BiCMOS NAND gate (Fig. 5.12) with Wp=30um, Wn1=26uM, Wn2=20uM, CMOS NAND gate with Wp=30um, Wn=26uM. All device lengths 1.5uM.

The BiCMOS circuit also benefits A.C. power dissipation compared to pure CMOS under certain heavy load driving conditions. Normally CMOS power calculations are based simply on the charging and discharging of the parasitic and load capacitances and are naturally a function of operating frequency, thus are given in terms of uW/MHz. For extremely large capacitive loads the required buffering of devices may introduce parasitic capacitance which substantially contributes to the overall power dissipation. A second aspect of power dissipation becomes significant for mismatched or parasitic RC limited signals that transition slowly such that the gate being driven slews through a midthreshold region where both devices are on and significant wasted "through" current results. The BiCMOS circuit, as a result of better load driving capability, tends to drive subsequent stages rapidly through this midthreshold region and thus the overall power dissipation is reduced because of diminished "through" current (Fig. 5.9). For power sensitive applications however, parasitic interconnect resistance may introduce RC limited signals at the input of a BiCMOS circuit which, due to the higher transconductance bipolar devices, will conduct even greater "through" current thereby resulting in greater power dissipation. If power is of primary concern, simulations including parasitic capacitance and resistance are appropriate; contributions to power dissipation from both the capacitive switching and input edge rate produced "through" current should be determined.

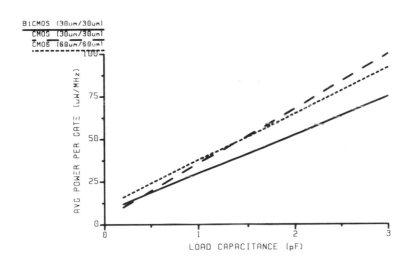

Figure 5.9 Average Power Per Gate vs. Load Capacitance
BiCMOS and CMOS gate power dissipation. BiCMOS gate with Wp=30um, Wn1=30uM, Wn2=20uM, CMOS gate with Wp=30um, Wn=30uM and double buffered CMOS gate with Wp=60um, Wn=60uM. (from [3]).

5.1.3 CMOS Delay Analysis

The propagation delay of the simple CMOS inverter may be approximated by treating all parasitic (voltage dependent) capacitances as a single constant lumped capacitance, C_{OUT}, at the output. Defining the propagation delay time as the 50% point at the input $((V_{OH} - V_{OL})/2)$ to the 50% point at the output, the average current available to charge (discharge) this capacitive load is used to calculate the delay as:

$$T_d = C_{OUT} \Delta V_{OUT} / I_{AVG}$$

The output must swing from its V_{OH} level (V_{DD}) to the midthreshold level ($1/2 V_{DD}$) so that:

$$T_d = C_{OUT} (1/2 V_{DD}) / I_{AVG}$$

The average current is approximated by averaging the initial current when the input is at midthreshold and the output is at V_{OH} with the current when the input is at V_{DD} and the output has fallen to the midthreshold level. Initially the MOS device is saturated so that the drain current is given by:

$$I_D = 1/2 \; \mu \bullet C_{ox} W/L \left[(V_{gs} - V_t)^2 \right]$$

Substituting $1/2 V_{DD}$ for the gate to source voltage at the 50% input point:

$$I_D = 1/2 \; \mu \bullet C_{ox} W/L \left[(1/2 V_{DD} - V_t)^2 \right]$$

At the later time $V_{gs} = 2 V_{DD}$ and $V_{ds} = 1/2 V_{DD}$ the device is in the linear region with drain current given by:

$$I_D = 1/2 \; \mu \bullet C_{ox} \; W/L \left[2(V_{gs} - V_t)V_{ds} - V_{ds}^2 \right] \; or,$$

$$I_D = 1/2 \; \mu \bullet C_{ox} \; W/L \left[2(V_{DD} - V_T)(1/2 V_{DD}) - (1/2 V_{DD})^2 \right]$$

The average current then is:

$$I_{AVG} = 1/4 \, \mu \bullet C_{ox} \; W/L \; [\; V_{DD}^2 + V_T^2 - 2V_{DD}V_T \;]$$

Thus yielding for the delay time:

$$T_d = 2C_{OUT} V_{DD} / \mu \bullet C_{ox} \; W/L \; [\; V_{DD}^2 + V_T^2 - 2V_t]$$

It is useful to define a parameter, $R_{EFF} = \Delta V_{OUT}/I_{avg}$, which has the units of resistance so that for given device parameters and output load, C_{OUT}, the delay is simply: $T_d = R_{EFF} \bullet C_{OUT}$.

5.1.4 Staged Buffer Delay Analysis

The gate and parasitic capacitances of certain high fanout nets or outputs can easily total into the several picoFarads. In the case of clock signals for large synchronous circuits it is typically required that the active clock edges occur simultaneously throughout the chip. This precludes the use of a branched tree style of signal distribution because of the variation of inverter delays across the die. Thus large buffers are required to switch a given load capacitance within a specified edge rate (rise and fall times are here specified from the 10% point (of $V_{OH} - V_{OL}$) to the 90% point). A minimum sized internal device is not capable of driving the large gate capacitance of the output device directly, thus several stages of buffering are required before eventually the desired sized device is achieved. Typically staging is accomplished by making each subsequent driver's width larger by some constant factor, thus increasing the drive capability but at the same time increasing the gate input capacitance by this same factor. Given known load capacitance, C_{LOAD}, internal device geometry, W_{INT}, and required switching speed T_D; a typical design problem needs to resolve the optimum number of stages, N; the optimum scaling factor, K; and output device size, W_{OUT}. This scaling factor K may also be interpreted as the optimum fanout factor for design problems such as exist in gate arrays where all device sizes are the same (a more general analysis including many second order effects is presented by Glaser et. al. [2]).

The staged buffer of Fig. 5.10 illustrates the problem with N being the total number of stages, and C_1 representing the input load capacitance of inverter I_1 and so on for each inverter I_i with characteristic input capacitance C_i.

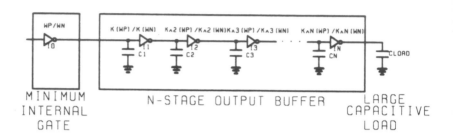

Figure 5.10 N Stage CMOS Driver

Neglecting interconnect and source drain capacitance and concentrating on the dominant gate input capacitance the assumption of constant scaling leads to:

$$C_{i+1} = K \bullet C_i$$

(assuming constant minimum lengths throughout) and

$$C_{LOAD} = K \bullet C_N \text{ and } C_0 = C_{INT}$$

thus it is possible to solve for the scaling factor K in terms of C_{INT} and C_{LOAD}:

$$K = (C_{LOAD}/C_{INT})^{1/N}$$

The total delay through such a staged CMOS driver may be expressed as:

$$T_d = \sum_{i=1,N} t_i = \sum_{i=1,N} R_i \bullet C_{i+1}$$

Where R_i is the R_{EFF} as defined above for the i'th inverter. Since both R_i and C_i scale roughly with the width by the constant factor K then each of the t_i are the same such that:

$$t_i = t_0 = R_{INT} \bullet (K \bullet C_{INT})$$

and the total delay is given:

$$T_d = N \bullet R_{INT} \bullet (K \bullet C_{INT})$$

minimizing the total delay with respect to the number of stages yields:

$$d/_{dN}(\ln T_d) = d/_{dN}\{ \ln(R_{INT} \bullet C_{INT}) + \ln N + N^{-1} \bullet \ln(C_{LOAD}/C_{INT}) \}$$

$$1/T_d \; d/_{dN} T_d = 1/N - N^{-2} \bullet \ln(C_{LOAD}/C_{INT})$$

solving for the derivative and equating to zero, the optimum number of stages and the multiplying factor K are found:

$$N = \ln(C_{LOAD}/C_{MIN})$$

$$K = (C_{LOAD}/C_{INT})^{[\ln(C_{LOAD}/C_{INT})]} = e.$$

Based on this simplified analysis the required number of buffering stages increases logarithmically with load capacitance while the constant scale factor should be set near the value 2.7. Accounting for junction and overlap capacitance a factor of 3 to 3.3 is a more appropriate value [10]. A practical example illustrates the above design methodology. Suppose an internal 50MHz ring oscillator signal (a period of 20ns) is used as an internal clock. It is required that the clock buffer be able to switch an 2pF load capacitance with the clock edge rise and fall times not to exceed 2ns. To determine the number of stages of buffering and device sizes for the staged buffer required assume the ring oscillator inverters are sized with W_P=10um, L_P=1.5um, W_N=7um, and L_N=1.5um. In order to drive the heavy load the ring oscillator inverters should be buffered by N stages, with N given by:

$N = \ln(C_{LOAD}/C_{OX} \bullet AREA).$

Assuming a 225 angstrom gate oxide and an area of 25.5um^2 the optimal number of stages is four. The devices should be sized such that W_{P1} =27um, W_{N1} =19um, W_{P2} =74um, W_{N2} =52um, W_{P3} =200um, W_{N3} =140um, W_{P4} =544um, and W_{N2} =380um, with all lengths assumed L_P =1.5um, and L_N =1.5um. For such a staged CMOS inverter the typical propagation delay is T_P =1.9ns and the total area consumed by the devices is approximately 16,600um^2 (25.7mils^2). In order to drive the same load capacitance with a BiCMOS buffer, bipolar devices (with knee currents of 10mA) provide sufficient load driving capability to use a single stage buffer. The corresponding propagation delay for the single stage buffer is T_P =1.4ns and the area consumed is 2500um^2 (3.9mils^2). The CMOS to BiCMOS area ratio is 6.6 for this particular load and device technology. This area ratio naturally decreases as MOS features shrink, becoming about 2.5 at a 1.2uM CMOS technology although a greater area ratio of 4.0 results if worst case parameters are taken into account.. This relative area ratio becomes even larger as the load capacitance increases beyond the 2pF considered in this example. Overall this suggest a more meaningful area comparison between CMOS and BiCMOS custom design must account for the increased drive capability of the BiCMOS circuits; realizing that logic intensive, low load driving circuitry is implemented in pure CMOS.

5.1.5 BiCMOS Delay Analysis

For the BiCMOS circuit the propagation delay again depends basically on the current available to charge the output load. The analysis is understandably complicated by the bipolar devices which are charge controlled rather than voltage controlled as the MOS devices. In order to solve analytically this requires that careful attention be paid to the surplus base charge which controls the collector current. Several analytical delay equations have been recently proposed [11,12] which may be used for preliminary calculations. The work of Roseel et. al. [11] presents the most practical analytic delay equations to date for the circuit of Fig. 5.3b. The authors suggest three regions of gate operation: low-level injection, high level injection, and R_C dominated conditions; and have put forward delay equations for each. If sufficiently large, the bipolar devices will operate ideally with the collector current given as: $I_c = \beta I_b$. Under conditions of low collector doping the device may saturate as a result of a high collector resistance R_C. The collector resistance determines how much collector current the device will support before the intrinsic base collector junction becomes forward biased causing the device to saturate (device layout, an n+ buried layer, and a deep collector implant are all effective at minimizing R_C). A collector resistance of 100 ohms conducting transient current of just 1mA will induce voltage drops of 100mV between the extrinsic collector contact and the intrinsic collector base junction. In practice currents in excess of 5mA are seen so that saturation frequently occurs. Under such conditions the delay is RC limited and the current will increase only logarithmically with time constant given as $R_C C_{OUT}$.

$$T_d = (V_{BEON} C1)/I_D + Tsat + R_C C_{OUT} \ln[(Vdd-Vs) \ /(Vdd-Vsat)]$$

where,

$$Tsat = C_{OUT} \{[(R_C + {}^{TF}/I_D)^2 + (2TF2 + VDD)/C_{OUT}]^{1/2} - R_C - {}^{TF}/C_{OUT}\}$$

$$Vsat = I_D Tsat^2/2C_{OUT} TF + {}^{I_D Tsat}/C_{OUT}$$

where C1 is the average capacitance at the base of the bipolar transistor (including MOS contributed overlap, junction, and gate capacitance and bipolar junction capacitances); I_D the average MOS current; Tsat the time to bipolar saturation; Vsat the output voltage at which the bipolar transistor first saturates; and Vs the gate switching threshold.

Low collector doping will also cause the bipolar device to enter high level conditions in which case the collector current will be drastically diminished from its nominal value. The relevant parameter is the knee current, I_K defined in terms of Gummel–Poon model parameters as $I_K \equiv Q_{B0}/TF$. Practically the knee is found as the current at which the forward current gain β is reduced to 50% of its maximum value as a result of base pushout at the collector (Kirk Effect) and high level injection at the base–emitter junction. Under high level conditions the log of the collector current is no longer proportional qV_{BE}/KT but rather to $qV_{BE}/2KT$ [13]. Under such conditions Roseel et. al. approximate the delay as:

$$T_d = (V_{BEON} C1)/I_D + (AC_{OUT} Vs)/(I_D I_K)^{1/2}$$

where A is an empirical constant.

Both I_K and R_C scale with area, thus there is an area trade-off and the application will determine whether operating the devices in these non-ideal regions is appropriate.

5.2 Basic BiCMOS Circuit Design

Although the above analysis leads to an important intuitive understanding of the factors affecting the dynamic performance of BiCMOS circuits, an *a priori* knowledge of geometry dependent bipolar and MOS device information is required. The typical design challenge on the other hand, consists of choosing rational first pass device geometries followed by optimization. In reality MOS and bipolar dimensions determine parasitic capacitances which play a key role in BiCMOS circuit switching speed; and optimization requires extensive device modeling and circuit simulation (such as SPICE) in order to account for the complex dynamic processes. Thus although no closed form analytical solution can lead directly to optimized device dimensions a practical design methodology for BiCMOS circuits can be formulated in the case where a large output capacitance dominates the delay, much as was done for CMOS. To achieve such first pass device dimensions, determine the average current required in order to accomplish the desired switching delay using a simple $T_d = C_{OUT}(\Delta V)/I_{AVG}$ approach. This I_{AVG} then must be carefully examined with respect to the high level injection and saturation effects described above. Given the bipolar device characteristics required the bipolar emitter area can then be calculated. These bipolar device dimensions, in turn, determine the MOS device dimensions from both a base current and dynamic capacitive load driving capability standpoint. The latter is the dominant requirement for an A.C. analysis and standard MOS scaling suggests that MOS devices with gate capacitances about 1/3 the equivalent bipolar input capacitances are required. The ratio of PMOS to NMOS geometries is then chosen so as to set the switching threshold (typically at $1/2V_{DD}$) keeping in mind that NMOS gate to source voltage is diminished by a V_{BE}. At this point transient simulations should be performed and base, collector, and source/drain currents, all voltages, and relevant capacitive charge storage mechanisms reviewed. It is important to vary sizes of both double decode and active base discharge NMOS devices independently with an eye toward overall dynamic response as well as transient through-current. If possible Response Surface Methodology [14] and Monte Carlo analysis should be applied at this stage to yield more robust design. In addition to device dimension optimization valuable feedback for processing should occur with the development of design imposed requirements for device parameters.

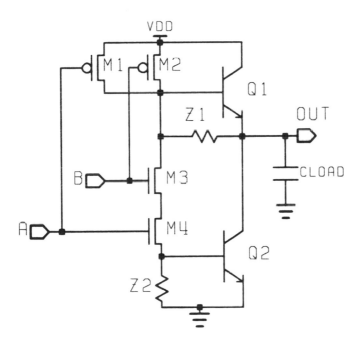

Figure 5.11 BiCMOS NAND Gate

The basic operation of the BiCMOS NAND gate circuit (Fig. 5.11) is similar to that of the inverter described previously and can be simply understood by considering the MOS devices first and realizing that the bipolar devices function strictly as output buffers.

For the case of a low logic level at either input A or B one of the series NMOS devices M3 or M4 is off and no current is supplied to the base of Q2. One of the two PMOS devices, on the other hand, is on and there exists a current path from VDD to the base of Q1. Thus Q1 will turn on and act as an emitter follower, sourcing output current to the load. Similar operation occurs when both inputs are high in which case both M1 and M2 are off and M3 and M4 are on supplying base current to Q2 which discharges the load. It is important to note that devices M3 and M4 will be able to draw current from the base of Q1 thus rapidly turning this device off. The resistive devices shown are necessary for proper operation of the circuit but typically are implemented as active (MOS) devices rather than diffused resistors. The "resistor" Z1 is necessary in order to achieve a low level at the output since without it there is no D.C. base current path to Q2. The second resistor Z2 functions to rapidly discharge the base of Q2 during a low to high transition at the output. It is important to note that neglecting reverse biased leakage currents the static power consumption of the circuit is negligible. Considering the D.C. logic levels provided by the circuit the V_{OH} level corresponds to that of CMOS circuits and effectively the output pulls to the rail. The V_{OL} level likewise can pull to ground via the series path through Z1, M3, M4, and Z2.

Clearly the implementation of more practical circuits follows that of the inverter so that circuits such as the one in Fig. 5.12 are used, and as pointed out above, do not possess the full output voltage swing of standard CMOS gates. One advantage of this type of circuit is that during a falling output transition the active device M7 turns off as a result of M3 and M4 discharging the base of Q1 and pulling the gate voltage of M7 low. This means that all the current through M5 and M6 will be provided as base current to Q2 as opposed to the circuit of Fig. 5.11 in which some of the current is shunted through Z2 directly to ground. Similarly a rising output may shunt current through Z1 to pull up the load directly and thereby lose some of the current gain available by driving all the current into the base of Q1. This occurs of course at the expense of active device count in order to perform the NMOS double decode logic.

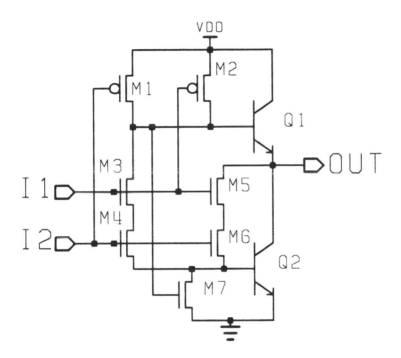

Figure 5.12 Practical BiCMOS NAND

The extension to more complex circuits follows closely standard CMOS design techniques, although special consideration must be taken for certain types of circuits. For example note the additional MOS device M7 necessary in the implementation of a BiCMOS tristate inverter as shown in (Fig. 5.13). This device is necessary when the enable signal goes low in order to actively discharge the base of Q2 so as to tristate the output.

The buffering of the MOS devices from large output loads allows smaller devices than would otherwise be needed and uses the current controlled current source bipolar device instead of the relatively low transconductance MOS devices to source output switching current. This is important since the current drive of a MOS device in the linear region is proportional to the drain source voltage across it and thus the switching delay of a CMOS gate is directly affected by decreases in the power supply voltage. Although the MOS supplied base drive is degraded by reduced supply voltages, the bipolar devices moderate load charging delays by the current gain β which is frequently in the 80–100 range.

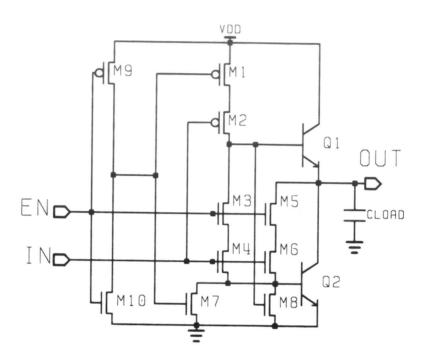

Figure 5.13 BiCMOS Tristate Inverter

5.3 BiCMOS ASIC Applications

The application specific market has been especially receptive of BiCMOS technology based on several conditions arising from the gate array and standard cell products [15,16,17,18]. Among these are large capacitive loads resulting from inherent high fanout, long interconnect paths associated with high gate count arrays, and the relative ease of implementing BiCMOS circuits into gate array architectures. As an example of the capacitive loads that might be encountered consider a CMOS inverter driving a fanout out of ten with an assumed input capacitance of .12 pF/gate and metal capacitance of .004 pF/mil metal routing. Using an estimate of 50 mils routing metal the total capacitance that the load must drive is approximately: C_{load} =

$10 \cdot (0.12pF + 50 \cdot 0.004pF) = 3.2pF$. Thus a natural limitation for CMOS gate array performance results from the long delays associated with driving these large capacitive loads. Additionally for the bulk of digital gate array designs the requirements of synchronous design exacerbate the problems outlined above. Such designs frequently use one or more common bus structures with inherent high fanout and loading. A sample of the typical number of gates driven (fanout) by each output node for options of both high and low bus usage (Fig. 5.14), illustrates the high fanouts in modern VLSI arrays.

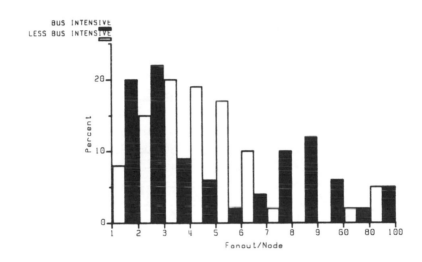

Figure 5.14 VLSI GATE ARRAY FANOUT

Typical gate array fanout per internal output for options of low and high internal bus utilization (from [14]).

Additionally it is apparent that as array size increases the capacitive fanout median increases and a bimodal distribution is observed (Fig. 5.14). This is a direct consequence of Rent's Rule. Examination of critical delay paths in typical gate array applications then indicated that over 60% of the functions in the critical path benefit from the buffered CMOS cell; the improvement being a factor of ~2. The penalty for this improved performance is a slight area penalty: 20% for a 2-input Nand and 5% for an adder. Power dissipation however, is significantly lower for all capacitive loading greater than the cross–over point.

Also apparent is the small but significant number of nodes with fanouts in the 60–100 range which clearly are associated with clock drivers. In large designs a single clock signal may be required by a hundred or more inputs, and using separate buffers as is often done introduces a clock skew problem which the designer must account for when choosing a maximum clock frequency for the system. The clock distribution solution offered by high drive BiCMOS circuits allows paralleling of buffers with a much higher drive capability than similarly available CMOS clock buffers.

The BiCMOS buffer circuits provides the capability of tighter clocking skews by a factor of 2, over the entire array. In a large array CMOS clock skews can be 1ns or greater compared to 500ps for BiCMOS. The improvement stems from the higher drive per unit area of the BiCMOS gate which makes for a very shallow clocking tree. A shallower clocking tree then results in a reduction in the effect of process variability on skew, as variation in each stage is additive. The net result can be up to a 2X improvement in system performance.

Application specific arrays with embedded architectures are finding broader acceptance and lend themselves to a merged technology. Three distinct classes have emerged: RAM [19], ECL core [20], and Analog [21]. A fourth utilizing CPU macrocells may also fill a niche. RAM arrays can be used for pipeline type options where data can be quickly fed to the BiCMOS internal logic. RAM arrays utilize a variety of circuit techniques to take advantage of CMOS, BiCMOS, and ECL. The basic cell is MOS which allows high packing density, low power dissipation. and level sensitivity. While a 4T cell is preferred because of area considerations (typically a factor of 1.5 –2), the extra process complexity (2–3 mask steps) involved with the 4T cell, and the improved stability, alpha hardness, and margins of the 6T cell, have led most manufacturers to adopt a 6T cell. The sense amps are ECL for fast read: typically a factor of 2 better than CMOS sense amps. BiCMOS address/decode sections exploit their high drive, low skew characteristics resulting in another factor of two improvement in speed. Arrays with embedded ECL sections accommodate very high speed critical delay paths and are especially useful for high speed clocks [20]. Analog sectors are a natural application of BiCMOS technology. Here it is possible to develop high speed, high accuracy data conversion devices. The analog functions combine with the digital interface around the array periphery. This is especially useful for DSP applications like custom filtering functions, high resolution display drivers, and modem interfaces.

The gate array market (which has proved to be much larger than the standard cell market among ASICS) has several features which make BiCMOS technology attractive. First BiCMOS circuits tend to balance the skews between NAND and NOR circuits better than standard CMOS gate array circuits. A CMOS gate array is distinguished by a regular pattern of fixed diffusions of both NMOS and PMOS devices which are then customized by, typically two or three layers of application specific metal interconnect to form the appropriate SSI gates and connections between them. The gate array vendor normally chooses a ratio of PMOS to NMOS widths for the diffusion base so as to optimize NAND gate performance.

This then inherently means that the MOS devices are not optimized for a NOR gate since now the PMOS devices are in series resulting in poor drive capability and therefore slower rise times compared to fall times. A comparison of the typical rising and falling propagation delay times of the BiCMOS and CMOS versions of a four input NOR gate (both implemented in 1.5uM technology, $Z_p = 30uM$, $Z_n = 26uM$, driving a fanout of five) illustrates the disparity:

BiCMOS: $T_{phl} = 0.44nS$, $T_{plh} = 1.08nS$
CMOS: $T_{phl} = 0.43nS$, $T_{plh} = 1.60nS$

Though neither is symmetric the BiCMOS gate's skew is considerably less than that of the CMOS gate. A second feature of gate arrays which lends itself well to BiCMOS circuits is the repetitive architecture, since the bipolar diffusions may use area otherwise wasted under power and ground metal busses and metal routing channels of many arrays. The newer channelless or "sea of gates" architectures have produced proposed BiCMOS strip architectures or annular ring architectures [16] since it has become apparent that a balance between BiCMOS circuits and pure CMOS circuits is desired by many designers. An example of one of the BiCMOS gate array products currently available is shown in Fig. 5.15.

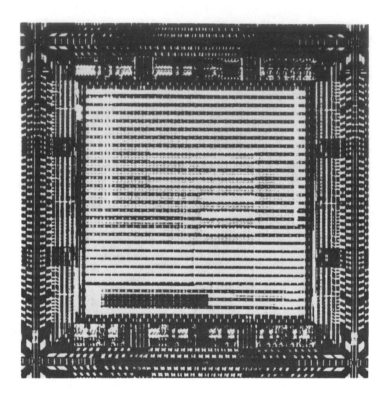

Figure 5.15 BiCMOS GATE ARRAY

An available ECL/TTL/CMOS compatible BiCMOS Gate Array with 1.5um features, 700ps typical delay (fanout=2, 2mm of metal) and up to 13,440 equivalent gates. Die photo courtesy of Applied MicroCircuits Corporation.

The development of the gate array market has occurred as system designers attempt to consolidate standard glue logic boards into single chip solutions. Large computer and communication systems may, however, use several different technologies which necessitate surrounding gate arrays or microprocessors and peripherals with glue logic translators to allow interconnection of TTL, CMOS, and ECL signals. BiCMOS circuits allow extremely elegant and flexible techniques to eliminate these translators and accomplish mixed system communication [19,22].

CMOS products have been relatively successful at accomplishing input circuits with TTL compatible thresholds but much less so with TTL compatible outputs. The simple approach of skewing device sizes toward the NMOS device (typically dimensions must be set such that $(W/L)_n = 5(W/L)_p$) has allowed input threshold voltages of CMOS circuits to be set at or near the nominal threshold voltage of an LSTTL input: 1.3V. This approach however results in the need for subsequent buffering in order to restore symmetry to rising and falling delays thereby adding to the overall delay of the input. Furthermore since the threshold of this type of input is much more variable than a true TTL input there is a greater potential for latching a false positive edge resulting from noise spikes on the VSS lines caused by the simultaneous switching of other outputs. Lastly this type of TTL input does not yield similar temperature and voltage characteristics. The availability of Schottky diodes and Schottky clamped bipolar devices, to a large degree, determines the design approach for BiCMOS TTL compatible circuits. Fortunately it appears that advanced submicron process phenomena may require some form of barrier metal which may, with proper process design, be incorporated to form the desired Schottky devices [23]. In this case the BiCMOS circuits can resemble very closely those used to design LSTTL circuits with the additional benefit that resistors with non−critical values may be replaced by active MOS devices resulting in better transient performance and lower D.C. power requirements.

The BiCMOS TTL output circuit (Fig. 5.16) uses a Schottky clamped darlington output and RTL phase splitter similar to modern FAST technology. The MOS transistors M3 and M5 are used to rapidly turn off pulldown transistor Q3 during low to high transitions at the output but do not provide a D.C. leakage path when the output is low. The MOS transistor M4 performs a similar function during high to low output transition discharging Q2 while providing A.C. base current to Q3. The static characteristics of this output circuit are very similar to that of standard TTL logic due to the similar nature of the circuitry. The TTL output low voltage V_{OLMAX} spec is normally 0.5V at an $I_{OL} = 8mA$ and is determined by the V_{CESAT} of the pulldown transistor Q3. This V_{CESAT} is in turn determined primarily by the parasitic collector resistance, R_C, which may be reduced by introducing an N+ buried layer, normally present in bipolar processes, to the BiCMOS flow. This requires additional photo and implant steps and an epitaxial growth step but the N+ buried layer can be used to reduce the parasitic MOS well resistances to provide the added benefit of virtual immunity to latchup.

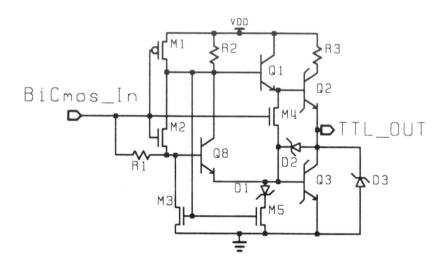

Figure 5.16 BiCMOS TTL Output

An important market for BiCMOS gate arrays has arisen to incorporate TTL circuits in systems required to drive the relatively low impedance of a backplane data bus. In order to achieve high data rates, transmission line effects such as reflections and ringing must be circumvented by using impedance matching termination of transmission lines. A common termination procedure uses a resistor divider termination providing an equivalent termination resistance of approximately 120Ω so as to match the characteristic impedance of the backplane [Fig 5.17]. The TTL open collector driver then must be able to sink 24mA (or 48mA in the case of termination at both ends) and still maintain a low voltage. Although commercially available CMOS gate arrays provide 24mA outputs they require extremely large devices and are much more efficiently implemented in BiCMOS.

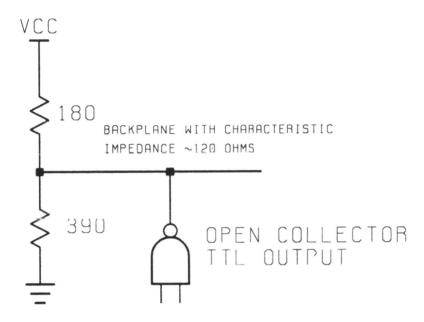

Figure 5.17 TTL Backplane Driver Requirements

The ability to interface CMOS or TTL directly with ECL circuitry is not simply done with MOS techniques and virtually impossible without the use of bipolar devices. The voltage level shifting, relatively small noise margin (1.05V 100K Family), and temperature sensitivity of ECL circuitry is not easily implemented with standard CMOS techniques and normally mixed ECL/CMOS systems require the use of dedicated translator circuits. The advent of BiCMOS circuitry enables the gate array user to implement mixed mode digital communication between ECL and TTL/CMOS circuits without the necessity of a host of discrete translator circuits.

Several mixed system configurations are possible in which the true ECL or TTL levels may be referenced to a non-typical supply (Fig. 5.18). The internal gate array is then powered off either VDD and VSS as supply voltages or VSS and VEE supply rails. The pseudo ECL configuration requires all external ECL circuitry to tie VSS and

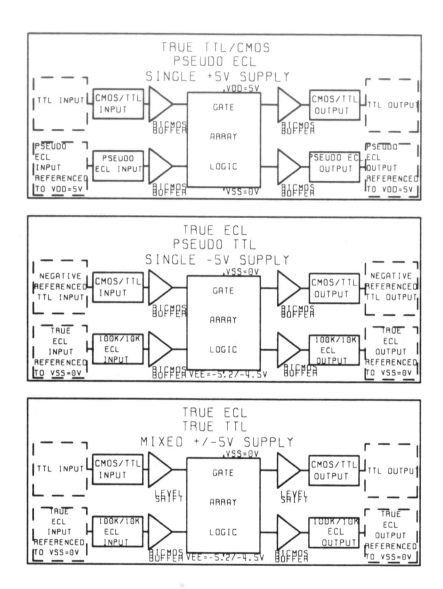

Figure 5.18 ECL Mixed Mode System Configurations

VSSO inputs to 5V and VEE to ground. This configuration suffers from the noise typically present on supply lines and this is the reason that ECL normally references all logic levels to the more solid ground reference. The pseudo TTL configuration on the other hand references TTL logic levels between ground and a negative VEE reference, however this tends to introduce TTL switching noise onto the ECL ground reference. Isolating this TTL output (and thus noise) to the VSSO supply tends to reduce the likelihood of inadvertently creating noise induced signals to ECL logic. Lastly the true ECL, true TTL configuration is the most robust from an interface standpoint but requires both dual supplies and dedicated internal translators to perform the necessary level shifting and logic level conversion.

The complicated temperature dependence of the Ic–Vbe characteristics of the Bipolar transistor [24] accounts in part for this difficulty to interface ECL circuitry with pure CMOS. The ingenious design of the 100K ECL NOR gate circuit (Fig. 5.19) illustrates how the negative temperature coefficient of the bipolar device (typically −1.5mV per degree Centigrade) may be circumvented to yield practically flat static output characteristics [25].

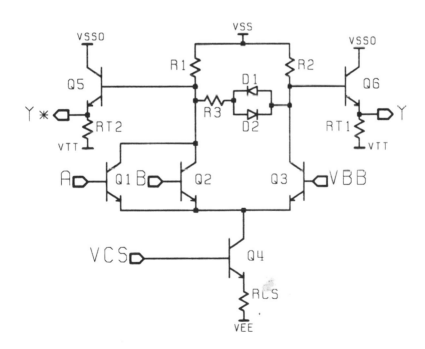

Figure 5.19 ECL 100K NOR Gate

Resistors R_1, R_2, and R_3 are identical valued resistors with value R. Neglecting base currents the static output parameters are found as follows. The output low voltage at Y occurs when the reference voltage V_{BB} is greater than either input voltage A or B and for differential voltages greater than a few hundred milliamps effectively all the switch current flows through Q_3. Thus

$$V_{OL} = V_{SS} - (2/3)I_{CS} \cdot R - (4/3)V_{BE}$$

The switch current $I_{CS} = (V_{CS} - V_{BE4}) / R_{CS}$ is designed such that $R/R_{CS} = 2$ and the current source reference voltage V_{CS} is bandgap stabilized and therefore temperature independent. Thus the temperature dependence of the current source offsets the temperature dependence of the V_{BE} term in the expression for V_{OL} thus yielding a temperature independent logic low level. The output high voltage at Y occurs when the reference voltage V_{BB} (nominally $V_{BB} = -1.3v$) is less than either input voltage A or B and again for differential voltages greater than a few hundred milliamps effectively zero current flows through Q_3. The output high voltage is found in terms of V_{OL} as:

$$V_{OH} = (V_{SS} - V_{OL} + 2V_{BE}) \cdot [R_1 / (R_1 + R_3)] - V_{BE} = 1/2V_{OL}$$

with V_{SS} equal to the zero volt ground reference. Thus given a temperature independent V_{OL} then V_{OH} will also have a zero temperature coefficient. The constant voltage V_{BB} is the bandgap generated temperature independent reference level designed midway between V_{OL} and V_{OH}.

In order to match the temperature and supply dependence of such ECL circuitry it is virtually mandatory to implement the interface logic with bipolar devices (a notable exception is GaAs MESFET I.C's which have gone to great lengths to attain ECL compatibility [26]). This ability, to integrate I/0 and high speed critical paths in ECL along with high density, low power CMOS and BiCMOS circuits, alone assures the long term viability of the technology. New hybrid ECL/MOS circuits are possible, although to date the area has apparently received little attention. One potential application is to eliminate duplicate "slave" bias networks required to supply the finite base current to the V_{BB} and V_{CS} references thereby eliminating the "slave" current.

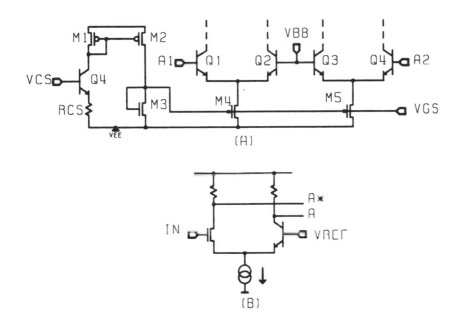

Figure 5.20 Merged ECL/MOS Circuits

One scheme to eliminate this current would replace the $(V_{CS} - V_{BE4})/R_{CS}$ current source of Fig. 5.19 with the MOS mirror circuit of Fig. 5.20A. The same current dependence needed for temperature compensation is generated and mirrored to other MOS current source devices. For current mirrors such as this careful consideration of layout would be required as a result of the inherently higher threshold voltage mismatch of MOS devices (common centroid geometry with the "donut" shaped MOS current devices positioned at the center of several surrounding ECL gates with drains chosen as the outside of the "donut" so as to increase output impedance). Since the MOS gates present a zero D.C. current load and correspondingly, there are no ohmic losses with which to contend; this configuration could enable a single bias network to be sufficient. A merged CMOS/bipolar logic (MCSL) circuit (Fig. 5.20B) has been announced which eliminates the emitter followers as in CML, however requires matching bipolar V_{BE}'s to MOS V_T's and requires D.C. bias current for reference levels [27].

5.4 Conclusions

Recent integrated circuit processing developments have allowed designers to implement hybrid BiCMOS circuits incorporating high Gm Bipolar devices to buffer CMOS outputs. A variety of such circuits have appeared on the market each imposing different performance or area tradeoffs. All however represent significant performance advantages over pure CMOS circuits because of better drive capability and less sensitivity to process, temperature, and supply variations. A BiCMOS delay analysis makes it clear that bipolar device characteristics (Rc, Tf, Ik ...) together with appropriately scaled MOS devices combine to determine overall gate delay. Lastly the incorporation of BiCMOS circuitry in the new semi-custom products indicates the versatility of the technology with regards to mixed TTL/ECL/CMOS circuits. Predicting the direction of semiconductor markets is much akin to predicting the stock market, and bipolar markets have prospered despite the prophecy of many CMOS supporters, and in turn the CMOS market has grown despite the early concerns of NMOS designers with regards to density and process complexity. And yet it is the design flexibility a merged BiCMOS technology provides that could well insure the continued growth of both technologies.

The author would like to thank Tony Alvarez for his valuable material and analytical input regarding BiCMOS design without which this chapter would not have been possible, and should also like to recognize the technical input of Doug Schucker.

References

1] David A. Hodges and Horace G. Jackson, "Analysis and Design of Digital Integrated Circuits", McGraw Hill, 1983.

2] Lance A. Glasser and Daniel W. Dobberpuhl, "The Design and Analysis of VLSI Circuits", Addison Wesley, 1985.

3] Motorola BiMOS Macrocell Array Design Manual, 1988.

4] Hiroshi Nakashiba, Kaszuyoshi Yamada, Tsutomo Hatano, Akira Denda, Norio Kusonose, Eigo Fuse, and Masakazu Sasaki, "A Subnanosecond Bi-CMOS Gate-Array Family", IEEE CICC, 1986.

5] Takahide Ikeda, Atsuo Watanabe, Yoji Nishio, Ikuro Masuda, Nobuo Tamba, Masanori Odaka, and Katsumi Ogiue, "High-Speed

BiCMOS Technology with a Buried Twin Well Structure", IEEE Transactions on Electron Devices, Vol. ED–34, No. 6, pp. 1304–1309, 1987.

6] AMCC Q14000 Series BiCMOS Logic Arrays Preliminary Device Specification, Macrocell Array Design Manual, 1988 .

7] Yoji Nishio, Fumio Murabayashi, Shoichi Kotoku, Atsuo Watanabe, Shoji Shukuri, Katsuhiro Shimohigashi, "A BiCMOS Logic Gate with Positive Feedback", IEEE ISSCC, Vol 32, pp. 116–117, 1989.

8] Paul R. Gray, and Robert G. Meyer, "Analysis and Design of Integrated Circuits", John Wiley, 1977.

9] A.R. Alvarez, J. Arreola, S.Y. Pai, K.N. Ratnakumar, "A Methodology for Worst–Case Design of BiCMOS Integrated Circuits", BCTM, PP. 172–175, 1988.

10] A.R. Alvarez, "BiCMOS vs. CMOS Buffers", Internal Correspondence, October, 1987.

11] G.P. Roseel, R.W. Dutton, K. Mayaram, and D.O. Pederson, "Delay Analysis for BiCMOS Drivers", BCTM, pp.220–222, 1988.

12] E.W. Greenich, K.L. McLaughlin, "Analysis and Characterization of BiCMOS for High Speed Digital Logic", IEEE J. Solid State Circuits, Vol. SC–23, no.2, pp.566–572, April 1988.

13] Richard S. Muller and Theodore I. Kamins, "Device Electronics for Integrated Circuits", John Wiley, pp. 244–287, 1977.

14] Jame McDonald, Rajnish Maini, Lou Spangler, and Skip Weed, "Response Surface Methodology; A Modeling tool for Integrated Circuit Designers", IEEE CICC, pp 13.4.1–13.4.4, 1988.

15] A.R. Alvarez, J. Teplik, D.W. Schucker, T. Hulseweh, H.B. Liang, M. Dydyk, I. Rahim, "Second Generation BiCMOS Gate Array Technology", Proceedings IEEE Bipolar Circuits and Technology Symposium, 1987.

16] "BiCMOS, Is it the Next Technology Driver?", Electronics, February 4, 1988.

17] A.R. Alvarez, D.W. Schucker, "BiCMOS Technology For Semi–Custom Integrated Circuits", Proceedings of the CICC, 1988.

18] Anthony Wong, Alex Hui, Eric Chan, Dan Wong, Steve Chan, Bill Carney, "A High Density BiCMOS Direct Drive Array", IEEE CICC, pp. 20.6.1–20.6.3, 1988.

19] P.S. Bennett, R.P. Dixon, F. Ormerod, "High Performance BIMOS Gate Arrays with Embedded Configurable Static Memory", CICC, 1987.

20] F. Ormerod, D.W. Schucker, K. Deierling, N. Salamina, "A Mixed Technology Gate Array with ECL and BiMOS Logic on a Single Chip", Symp. on VLSI Circuits, pp. 31–33, 1987.

21] Y. Kowase, "A BiCMOS Analog/Digital LSI with the Programmable 280bit SRAM", ISSCC, 1986, PP.170–173, 1985.

22] Yasuhiro Sugimoto, and Hiroyuki Hara, , "Bi-CMOS Interface Circuit In Mixed CMOS/TTL and ECL Use Environment", First International Symposium on BiCMOS, 1986.

23] S. P. Murarka, "Silicides for VLSI Applications", Academic Press, 1983.

24] Yannis P. Tsividis, "Accurate Analysis of Temperature Effects in Ic–Vbe Characteristics with Application to Bandgap Reference Sources",   IEEE JSSC, Vol SC–15, no. 6, pp. 1076–1083, December 1980.

25] Fairchild F100K ECL Data Book, pp. 2–7, 1982.

26] R. Eden, J. Clark, A. Fiedler, F. Lee, "VBB Feedback Approach for Achieving ECL Compatibility in GaAs ICs", Tech. Digest GaAs IC Symposium, pp. 123–127, 1986. Sources",   IEEE JSSC, Vol SC–15, no. 6, pp. 1076–1083, December 1980.

27] W. Heimsch, B. Hoffman, R.Krebs, E. Muellner, B. Pfaeffel, K. Ziemann, "Merged CMOS/Bipolar Current Switch Logic", IEEE ISSCC, Vol 32, pp. 112–113, 1989.

Chapter 6

BiCMOS Standard Memories

H. V. Tran, P. K. Fung, D.B. Scott and A. H. Shah,
(Texas Instruments Incorporated)

6.0 Introduction to BiCMOS Memory IC's

Over the past few years, memory performance has been the primary demonstration vehicle for BiCMOS technologies. Intrinsic gate delay, power dissipation and area have been regarded as the theoretical indications for technology performance and density. In a similar manner memory access time, memory power dissipation and memory size have been regarded as the practical indications of technology performance and density. Against this empirical yard stick BiCMOS technology has been found to produce memories with MOS-like power and density but with speeds and I/O interfaces which one normally attributes to bipolar memories.

By far the most common use of BiCMOS in memories is to start with the MOS memory cell and to surround it with a merged technology periphery. The merged bipolar and CMOS memory cell [6.1] is a notable exception to this trend. As the design techniques mature, the bipolar devices tend to get incorporated into more and more of the critical path delays.

Bipolar devices impact overall SRAM performance in a number of ways. The most obvious example is the use of high speed ECL circuits at the input and the output of the memories. Another example is the greater current carrying capability which allows bipolar devices to be used for on-chip buffer applications. This is particularly useful when the on-chip buffers must drive very heavy capacitively loaded lines. In addition to the above, bipolar devices are preferred for differential amplifiers because of their superior matching capability. In fact, all BiCMOS SRAM's use the bipolar devices for signal sensing.

In the following chapter the architecture and functional blocks of memories are described. Following this the specific implementation details of each type of memory circuit is discussed. This includes the circuits used throughout the chip as the address signal gets decoded to a specific memory cell and the signal path from the memory cell to the final chip output. Finally a discussion of the future outlook for BiCMOS is given.

6.1 Architecture and Functional blocks

Much of the work in BiCMOS technology has been focused on the SRAM. There are several reasons for this attention to static RAMs. First of all, it is absolutely necessary to have a memory device as a process technology driver (as opposed to manufacturing technology driver) for new and somewhat revolutionary process development. A memory device due to its regularity in design and layout serves as an excellent vehicle to "debug" the processes. Secondly, an SRAM is considerably simpler from design and debug standpoint. The last but not the least important reason is that the speed requirements of high performance computer systems can not be satisfied by the conventional approaches with DRAM's. The cache systems with static RAM's have become very popular because of the simpler control circuits and the high speeds possible in SRAMs.

6.1.1 Static Random Access Memory:

The static RAM, ever since the first inception, has not changed much in the basic operation. However, a lot of evolutionary enhancements have occurred over the years in the quest for higher speed, higher density, and lower power. Some of these enhancements can be classified as revolutionary. In addition to process technology innovations, many circuit and architecture innovations have taken place, including the basic memory cell structure.

Functional Description. The early versions of SRAMs were primarily singlebit output devices. The cells were relatively large and they served the very basic function of data storage in the computer systems. The applications of SRAM's have proliferated since then, and specific chip functions and architectures have evolved with them. We will review the SRAM at the functional level in this section and address the array architecture in the following section.

Fig. 6.1 shows a functional block diagram of a basic SRAM. These functional blocks are totally technology independent. However, how efficiently these functions are carried out depends very much on the circuit design techniques and the process technology. Even though with scaling and with innovative circuit designs a CMOS SRAM has gained considerable speed advantage over its predecessors, it has yet to match the speed of a bipolar SRAM. The BiCMOS process provides high performance bipolar transistors to a CMOS circuit designer and high density CMOS devices to a bipolar circuit designer. There are ten major blocks in a SRAM as shown in Fig. 6.1.

In the simplest form BiCMOS technology provides CMOS SRAMs a gateway to the ECL interface world. Even though there are CMOS circuits that provide direct ECL interface [6.2], they require special care in order to meet the bipolar speeds. The bipolar devices also provide high capacitive load drive

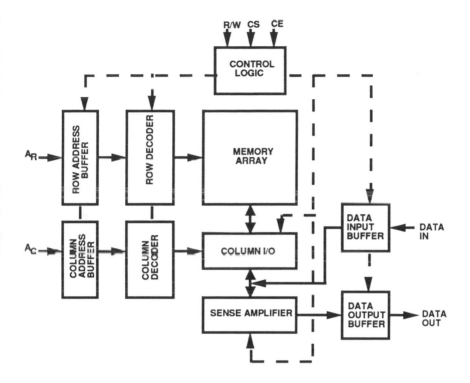

Figure 6.1 Functional Blocks of a typical SRAM.

capability that is unmatched by state-of-the-art scaled CMOS technology. Moreover, further scaling of the CMOS technology is currently too expensive for the speed advantages it can offer.

While BiCMOS is the ideal process technology to meet speed/density/power challenges of the next generation of SRAMs, it promises to pose a few challenges of its own, especially to the circuit designers. The inclusion of bipolar transistors and ECL requires a thorough consideration of reference voltage generation, the voltage translators, and the power and ground distribution, to name a few. The details of some of these functional circuits will be addressed in later sections of this chapter.

Array architecture. The memory array or the "core" is the main component of any SRAM. This core interfaces with the row selection circuit (row decoders) on one side and the column decoders as well as sense amplifiers on the other side. Fig. 6.2 shows a very basic array architecture for a typical SRAM using a divided word line[6.3]. The propagation delay through the array can be broken down into three major components: word line delay, bit line delay, and the sensing/amplification delay. The delay through the array is the major speed limiting component in a typical SRAM. The word line and bit line delays are primarily governed by their respective RC time constants and the sense amplifier speed is governed by the circuit configuration and pitch-limited layout constraints. Various innovative circuit and architecture ideas have been implemented to reduce these component delays. The divided word line approach is noteworthy because it reduces the word line time constant drastically by reducing the effective length of the active word line segment. Moreover, it also reduces the number of memory cells that are active simultaneously. Thus the active power dissipation is reduced. Almost all of the approaches to reduce the word line delay aim to reduce the effective word line length since the RC delay is approximately proportional to the inverse square of the word line length.

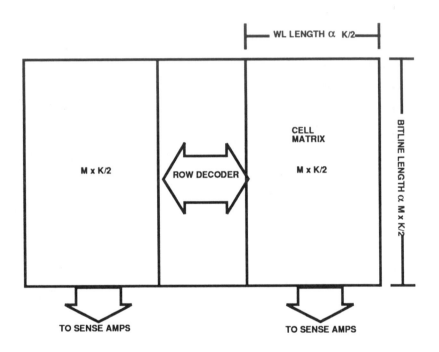

Figure 6.2 Basic array architecture using the divided word line approach.

The bit line delay in general is secondary to the word line delay because of the limited voltage swing. There have been new memory cells described that limit the word line swing to reduce the delay [6.1]. However, the push for higher density limits the cell choice to conventional cells such as full CMOS 6 transistor(6T) or 4 transistor, 2 resistor (4T-2R). While it is necessary to reduce the total memory cell area, its linear dimensions are constrained to a certain extent by the package and architecture considerations.

6.1.2 Dynamic Random Access Memory:

Although DRAM's have dominated the large and relatively slow memory systems, their performance continues to improve. BiCMOS DRAMs have not gained as much popularity as the BiCMOS SRAMs. DRAMs are predominantly driven by density and cost whereas a considerable segment of the SRAM market is driven by performance and density. However, DRAM applications have proliferated significantly over the last three to four generations. Some of these applications require very high performance along with the very high density of DRAMs. BiCMOS technology is well suited for such devices [6.4,6.5].

Functional Description. Functionally there is not much difference between an SRAM and a DRAM. Fig. 6.3 shows a typical functional block diagram of a DRAM. However, unlike SRAMs, due to the emphasis on memory system density, DRAMs have traditionally shared the row and column address circuits to reduce the pin count, and thus the package size. Even though this multiplexing scheme adversely impacts the performance, it provides significant density improvements and since the density is of paramount importance, the 1-transistor DRAM cell has been very difficult to replace.

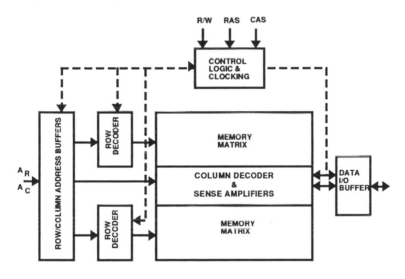

Figure 6.3 Block diagram of a typical DRAM.

Another difference between these memory types has been in the circuit design methodology. The SRAMs have traditionally employed static circuits in the periphery whereas DRAMs use dynamic circuits. However, even this major difference is beginning to disappear. The CMOS DRAM uses static circuits in the column section and the SRAM's are moving towards the clocked circuits.

Array Architecture. The most significant difference between the array architectures of SRAMs and DRAMs stems from the memory cell structure. Since the DRAM memory cell stores the charge on a capacitor, every unselected cell in a selected row has to be sensed and restored along with the selected cell. Consequently, one sense amplifier has to be provided for each column (or cell) in a row.

The SRAMs have traditionally multiplexed sense amplifiers among several columns. Moreover, each cell in a DRAM has to be sensed and restored at refresh time intervals. This single requirement drives the array architecture in terms of number of rows and columns. Moreover, the bit line delay in a DRAM is not a negligible portion of the total sensing delay. Thus, the DRAMs speed is limited by the word line and bit line lengths which are primarily determined by the refresh time requirements. These delays are minimized by new process technologies and cell structures as well as circuit design techniques.

6.2 BiCMOS Implementation

The diversity of circuit applications for BiCMOS can best be illustrated by the examples of how BiCMOS circuits are applied. In the following, various implementations of BiCMOS circuits are considered. These implementations are considered in the rough order in which a signal propagates through a large memory during a read cycle. Circuits are discussed starting with the input buffer, all the way through the row address circuits to the memory cell itself, then back out of the RAM from the bit lines through the data lines and to the final chip output. Although the circuit descriptions have been primarily done from the SRAM point of view, DRAM implementations are discussed as appropriate.

6.3 Input Buffers

As the access time of BiCMOS SRAMs improves to near the speed of ECL SRAM's, the circuits which were not considered very important in pure CMOS SRAM designs become crucial when in the critical speed path of a BiCMOS SRAM. The input buffer is one of the circuits which needs more attention than in the past.

6.3.1 ECL Input Buffer

One of the most important advantages of the BiCMOS circuit is the ability to directly interface with the ECL logic input levels. The availability of the bipolar device allowed circuit designers to use the true ECL input and output buffer in the BiCMOS design. ECL I/O levels have been used in many high speed systems because ECL logic is significantly faster than the other standard digital logic family. The ability of the ECL I/O circuit to drive transmission lines has become the most important factor in larger and faster systems.

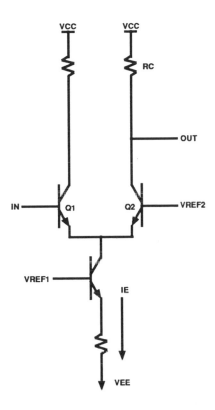

Figure 6.4 ECL input buffer.

A typical ECL input buffer is shown in Fig. 6.4. The ECL input signal is coupled directly to one input of a differential bipolar amplifier and a reference voltage is connected to the other input. The differential amplifier stage provides the voltage gain requirement for a small voltage swing of the input signal. The reference voltage level is set at the midpoint of the input logic swing to provide a precision detection of threshold between the logic '1' and '0' level. The small signal gain A_{dm} of the differential amplifier in Fig. 6.5 can be expressed as:

$$A_{dm} = \frac{\delta V_{out}}{V_{in} - V_{ref}} = -g_m R_c$$

and g_m is defined by

$$g_m = \frac{q I_E}{kT} = \frac{I_E}{V_t}$$

where V_t is often referred to as the thermal voltage,

$$V_t = \frac{kT}{q} = 25.9 \, mV \quad at \, 25 \, C$$

For I_E of 1mA and R_c of 1K, it can be seen that the small signal gain of the differential amplifier is about 40 and this is adequate for approximating the large signal amplification gain of the input buffer. Thus it is safe to assume the differential amplifier is functioning as a current steering element. The input signal steers the current I_E through transistor Q1 when the input voltage level is above the reference level and through transistor Q2 when the input voltage level is below the reference. The output voltage levels of the differential amplifier are:

$$Logic \, 1 = V_{cc}$$
$$Logic \, 0 = V_{cc} - I_E * R_c$$

The significant advantage of BiCMOS technology over the CMOS technology is underlined in the design and performance of the ECL input buffer. Major disadvantages of the MOS amplifier are the low transconductance (g_m) and mismatch transistor characteristics, which result in low noise margin and inadequate precision operation. The transconductance (g_m) of the MOS device operating in saturation region can be approximated as

$$g_m = \left(2 \mu C_o \frac{W}{L} I_d \right)^{1/2}$$

The comparison of bipolar and MOS transistor g_m's illustrated in Fig. 6.5 shows the size requirement of the MOS is quite large to have g_m equal to that of the bipolar transistor. A differential amplifier similar to the one in Fig. 6.4 designed with MOS transistors replacing the input bipolar transistors will suffer from switching speed degradation due to high capacitance loading, which results from the large layout area required to provide enough voltage gain and adequate noise margin. Reducing the size of the transistors and compensating the lower g_m with the higher current drive is only marginally effective since the g_m of MOS devices only increases as the square root of drain current (I_d). In addition to the lower g_m disadvantage, MOS transistors also have a greater mismatch characteristic between identically layed out transistors than the bipolar transistors. The mismatch results in an input offset voltage for the differential amplifier and thus further reduces the margin of the MOS input buffer.

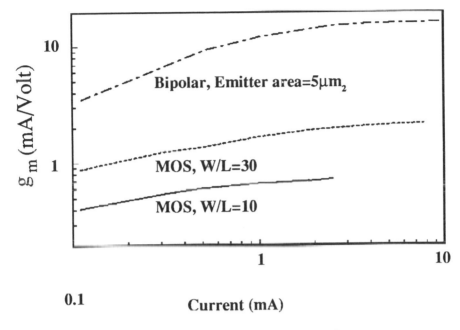

Figure 6.5 Comparison of bipolar and MOS transconductances.

In an attempt to force CMOS technology to accommodate the ECL interface I/O, many CMOS ECL input buffer designs have resorted to complex circuits as well as bias and compensation schemes [6.6,6.7]. This practice increases the susceptibility of the circuits to mismatch in devices, increases power consumption, and reduces noise margin. A simpler CMOS design [6.2], has recently been demonstrated to have the same class of performance as the BiCMOS design. However, high noise margin and low power consumption which are CMOS's inherent characteristics have been sacrificed. Also, non-standard multiple power supplies and complex level shifter external interfaces are required. In addition, the overall performance of the 0.5μ technology CMOS SRAM ECL I/O chip [6.8] barely outperforms a similar memory chip [6.9] which was two processing generations behind (1.3μ). This has greatly emphasized the advantage of the BiCMOS technology when it comes to the ECL I/O circuit design

A practical circuit for a BiCMOS SRAM ECL input buffer [6.10] which receives the external ECL level and converts to an internal CMOS level is shown in Fig. 6.6. The ECL signal is shifted down a total of two V_{be}'s before

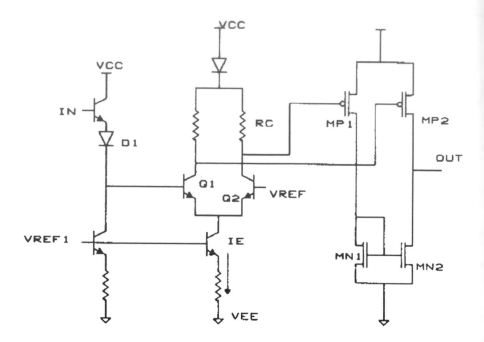

Figure 6.6 ECL input circuit with level translation.

passing the signal to the input of a differential amplifier. The outputs of the amplifier are also shifted down one diode drop from VCC and the high and low level swings are

$$V_{hi} = V_{cc} - V_{BE}$$
$$V_{lo} = V_{cc} - V_{BE} - I_E * R_C$$

The output of the differential amplifier which is about 0.8 volt is amplified again by the P channel transistor current mirror level shifter to a full CMOS signal. It can be seen that $V_{GS} - V_T$ or (V_{on}) of the P channel transistor in the current mirror pair, which is connected to V_{hi} level, will be

$$V_{on} = V_{hi} - V_{Tp} = V_{BE} - V_{Tp}$$

This P channel transistor will be off if V_{Tp} is higher than V_{BE}. The other P channel transistor of the current mirror pair, which is connected to V_{lo} level, will be turned on and its V_{on} is at

$$V_{on} = V_{lo} - V_{Tp} = V_{BE} + (I_E R_C) - V_{Tp}$$

Without the diode at the top of the differential amplifier, the V_{on} during the "on" state of the Pchannel would be one V_{BE} lower. However, in both designs, the V_{on} of the Pchannel will be less than 1.0v during the off state. Therefore faster switching is achieved by adding a simple diode at the top of differential amplifier. Diode D1 was added to the input stage to prevent Q1 and Q2 from saturating due to the additional one V_{BE} drop at the output of the differential amplifier. Node OUT will be pulled up to the VCC level by the P channel MP2 or pulled down to VEE by the N channel MN2 depending on the complementary output levels of the differential amplifier. V_{ref} sets the trip point of the input buffer as discussed before.

In many BiCMOS SRAMs designs [6.11,6.12], the ECL output levels from the differential amplifier are used to perform additional logic operations with ECL circuits before converting to CMOS signals. This approach enhances chip performance by operating the logic in the critical speed path with faster bipolar circuitry and at the same time operating the supporting circuitry in low power CMOS.

In addition to the advantage of directly interfacing with the ECL logic, BiCMOS technology also allows the ECL input buffer to be completely shut off and consume zero DC current. The ECL input buffer can feature a MOS current source instead of the standard bipolar current source, and this allows the current source to be turned off from a select signal. Thus high speed operation and low standby power can be combined in the same design.

6.3.2 TTL Input Buffer

Surprisingly, BiCMOS technology has not offered much advantage in the design of the TTL input buffer. Bipolar-only designs of the TTL input buffer have gone through numerous alterations and have not yet produced a major break through to improve the speed and power dissipation of the input buffer. Fig. 6.7 shows the standard design in bipolar. The low input impedance caused by the base current of the input transistor and the possibility of saturating the input transistor have been a nuisance for years. A Schottky diode placed at the base of the input transistor has helped to solve the saturation problem but input leakage and high power dissipation remain as disadvantages.

A TTL input buffer designed using a PNP transistor at the first stage can reduce the input current to the small base current of the input PNP transistor. However, switching speed will be compromised due to the slow speed of the PNP transistor compared to the higher speed NPN transistor and a DC current which flows through the buffer during circuit operation still remains. Furthermore, the PNP transistor is mostly available as a lateral device and its characteristics are not very controllable in processing.

The most common implementation currently is to use CMOS in the first stage as shown in Fig. 6.8.

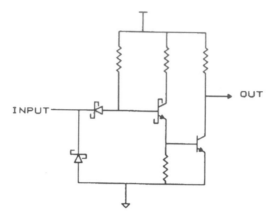

Figure 6.7 Schottky TTL input buffer design.

The CMOS gate in this design must resolve the input logic level above and below a threshold voltage of 1.4volt (midpoint of TTL levels). This low threshold level requires the Nchannel transistor size to be much larger than the P channel transistor and this severely affects the "pull-up" ability of the Pchannel transistor. However, the advantages of the high input impedance and compatibility with the main stream CMOS market have forced many BiCMOS designs to use the CMOS gate as the first stage input.

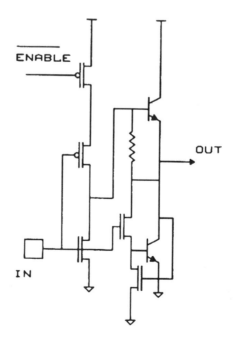

Figure 6.8 BiCMOS TTL input buffer design with CMOS input stage.

The TTL input buffer with its input trip point based on the switching threshold of a CMOS gate as used in the design of Fig. 6.8 has two major problems. First is the variations of input trip point across power supply, process and temperature. Second is the high power dissipation of the buffer's first stage inverter, which results from the small voltage swing of the TTL input level. Furthermore, for compatibility to the standby power requirements of the existing products, a CMOS input buffer often must be gated by an enabling signal for switching to and from the standby mode. These problems require compromises to be made between speed, power, yield and reliability, and thus tend to degrade the overall performance of the circuit. The transistor sizes of the buffer's first stage inverter are chosen such that the DC trip point is centered at the midpoint of the TTL input level (1.4 volts). However, as process, power supply and temperature fluctuate, the DC trip point deviates away from the midpoint and reduces the input signal margins as shown in Fig. 6.9. In addition, a significant current is flowing in the CMOS input buffer when its input is at TTL VIH level of 2.0 volts. This current is caused by the CMOS input buffer's first stage inverter which is partially in an on state.

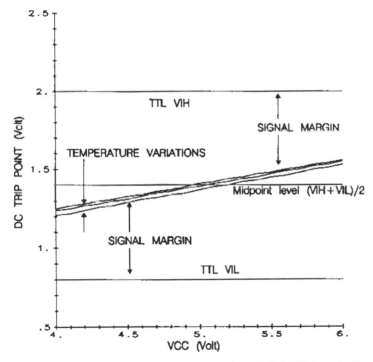

Figure 6.9 Trip Point vs Supply Voltage for a CMOS TTL input buffer.

A Threshold Control (TCON) BiCMOS TTL input buffer (Fig. 6.10) [6.13] can be used to eliminate the problems associated with the CMOS input buffer. The TCON input buffer uses a stable voltage reference source to establish the input trip point and minimize deviations caused by temperature, power supply and process variations. Low power dissipation is also achieved as a result of the buffer's circuit design.

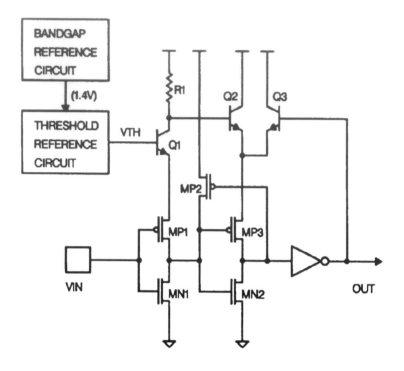

Figure 6.10 Threshold Control (TCON) BiCMOS TTL input buffer.

A BiCMOS band gap reference circuit, which is well known for its temperature stable output voltage, is used to provide a reference voltage of 1.4 volts to a Threshold Reference (TREF) circuit to establish a CMOS inverter trip point dependent V_{TH} signal [6.13]. This signal is connected to bipolar transistor Q1 to supply a regulated voltage level to the source of the Pchannel pull-up MP1. This voltage level is designed to keep the trip point of the TCON input buffer's first stage inverter at the midpoint of the TTL high and low levels.

The threshold reference voltage is chosen such that the voltage at the source of the Pchannel MP1 is always at or below a voltage level of:

$$V_{TH} - V_{BE} = V_{TTL(\ hi\ )} + V_{Tp}$$

where $V_{TTL(hi)} = 2.0v$ (TTL logic high level), and V_{Tp} is the V_T of the Pchannel transistor.

The buffer current is essentially zero when its input is held at a valid TTL level. This important feature allows the standby-to-active enabling signal to be omitted from the design of the input circuitry. The elimination of this enabling signal allows faster circuit operation because the output of the buffer is in a correct logic state during the standby period.

The TCON input buffer also utilizes a signal feed forward circuit technique to further enhance the buffer switching speed. Crowbar current of the first stage inverter is converted to a voltage glitch by resistor R1 atop the cascode transistor Q1. Bipolar transistor Q2 receives this voltage glitch and applies it to the source of the Pchannel transistor MP3 to help turn this transistor off during the transition of the input from high to low. This action compensates for the lower than VCC voltage at the output of the first inverter. The Pchannel transistor MP2 latches the second stage input up to VCC and eliminates crowbar current through transistors MP3 and MN2. When the input is switched from low to high, bipolar transistor Q3 stops the feed forward voltage glitch; and allows the transistor MP3 to retain its full gate to source potential during this transition operation.

6.3.3 Other Input Buffer Designs

Many systems utilize both ECL logic and TTL digital circuitry in a system board. Therefore the need for special input buffers which directly couple the signal from one type to another has arisen[6.14]. Fig. 6.11 shows a circuit used to receive ECL signals from an ECL device and converts those signals to a CMOS level. Power supply levels of the ECL device are 0v and -5v, whereas the power supply levels of

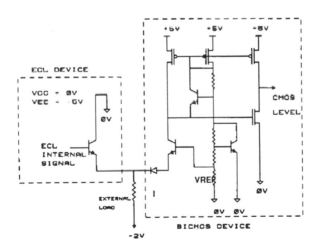

Figure 6.11 ECL to CMOS input buffer.

the BiCMOS chip are at +5 and 0v. Direct signal transfer is accomplished through the action of a psuedo differential amplifier comprised of the output transistor of the ECL chip, the load resistor and the input transistor of the BiCMOS chip. The base terminal of the ECL output transistor and the Vref signal are the two signal inputs and the load resistor serves as the current source. Current is steered to the output or to the input transistor depending on the voltage level at the base of the output transistor.

6.4 Level Conversion

Level conversion is a key issue in the use of mixed signal applications. Although ECL or TTL levels are often preferred for off-chip communication, zero power CMOS gates require the logic levels to coincide with the power supply voltages. For this reason high speed level conversion is essential to the power management of high speed integrated circuits.

6.4.1 ECL to CMOS

The major advantage of BiCMOS is high performance at high densities which is exactly what SRAMs capitalize on. Most of the BiCMOS SRAMs have a CMOS core and a BiCMOS periphery which receives either ECL or TTL signals. Therefore, it is necessary to interface the peripheral circuits with the core through level translators.

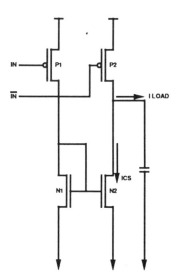

Figure 6.12 ECL to CMOS level converter.

For ECL I/O SRAMs, the ECL levels can be translated immediately after the ECL address buffer [6.10] or they can be translated after ECL predecoding [6.11,6.12]. The first method offers less power dissipation, but with a slight penalty in speed. The latter is faster at the expense of the extra power dissipated. Fig. 6.12 shows a basic MOS level translator [6.10] using an NMOS current mirror. The speed of the translator depends on the size of the MOS devices as well as the magnitude of the differential input signal. The current to the load is derived from the following equations:

$$I(load) = I(p2) - I(cs)$$

where $I(cs) = I(p1)$.

Thus

$$I(load) = I(p2) - I(p1).$$

The input signal must also have the correct level to be assured that the output of the translator swings to both power rails.

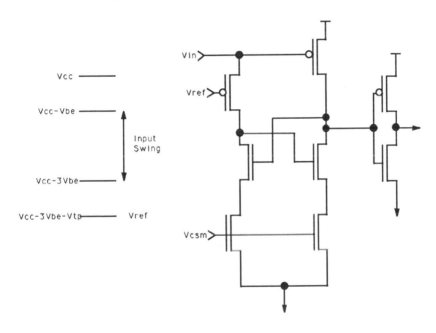

Figure 6.13 ECL to CMOS level converter using a voltage reference for single ended sensing after Kertis et al. [6.11]. Reprinted with permission of the publisher.
(© 1988 IEEE)

Fig. 6.13 [6.11] shows another ECL to CMOS level translator with a single-ended input plus two reference inputs. The ECL signal is tied to a PMOS gate and a PMOS source. Vref in Fig. 6.13 sets the trip point for the level translator while Vcsm determines the output level and speed. An optimum voltage exist for Vcsm whereby the speed is optimized while maintaining full level swing.

Most ECL-CMOS translators utilize some type of cross coupling latch or current mirror. Specific schemes are usually a variation of the basic translator.

6.4.2 CMOS to ECL

Fig. 6.14 shows a CMOS to ECL level translator. The trip point is determined by biasing Q2 so that its base is one V_{BE} lower than the base of Q1 when the PMOS transistor is on. When the PMOS is off the base of Q1 is clamped to VEE by the NMOS transistor, thus turning off Q1. Additional logic can be performed at the input of this translator to serve multiple signal high speed applications.

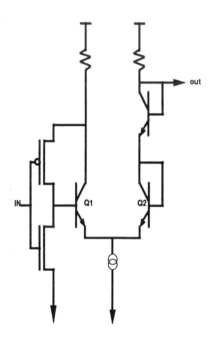

Figure 6.14 CMOS to ECL level translator.

6.4.3 BiCMOS to CMOS

It is often necessary to maintain signal levels at the supply rails to prevent excessive leakage. This is especially true when the signal drives many MOS gates. Signal levels at one V_{BE} from either supply rail can usually turn on most MOS devices weakly. If there are one thousand devices in parallel and each has a Vgs of 0.7 volts, a significant amount of leakage occurs.

During operation, and especially when driving heavy loads, BiCMOS gates swing to within a diode drop of the supply rails. Whenever a BiCMOS gate is used to drive either a word line or a series of MOS decode gates, the output of the

If larger devices are needed for heavier loads or for faster operation, other schemes can be used. In Fig. 6.15, the NMOS device is driven by a CMOS inverter which in turn is driven by the first stage. This scheme removes the loading of the extra device from the input. In addition, the device can be made larger because it has two stages of CMOS buffers.

Another possible scheme uses a latch technique whereby the BiCMOS output drives a CMOS inverter which then drives the parallel MOS device. Since the BiCMOS output is used to drive the CMOS inverter, there is no noticeable degradation in performance caused by the additional loading. Fig. 6.16 is an example of such a buffer.

6.5 Decoding

The major issue in the decoding of signals on a BiCMOS chip is the choice of logic type. This is important in that this choice varies depending on the functions on the chip. For ECL I/O RAM's it is advantageous to do a limited number of logical operations using ECL levels. However, as the number of gates involved increases, the advantage is with CMOS due to the zero inactive power. In the following sections the various considerations for decoding in BiCMOS are discussed for each type of logic level.

6.5.1 ECL Decode

Since BiCMOS was developed for performance advantages, it's evident that ECL logic levels should be used whenever possible. In BiCMOS SRAMs, ECL logic can be used in the circuit paths in and near the I/O for high speed without greatly increasing the power dissipation because of the limited number of circuits. Address buffers, predecoders, last stage sense amps, and output buffers are examples where ECl logic is effective. ECL logic is not used in paths such as decoders where the circuit is arrayed because the d.c. power dissipation would prove prohibitive.

The basic ECL logic function is a differential pair which has been shown in Fig. 6.4. A NOR/OR function can be implemented by adding transistors in parallel to one side of the differential pair. An added flexibility of ECL logic is the double-ended output of the gates. Thus the need to invert is avoided. Emitter-followers are added to ECL gates when driving large capacitive loads to insure the rising edge is as fast as the falling edge.

In using ECL logic gates, multiple inputs can be implemented per gate. Care, however, must be taken to limit the number of inputs when designing for high speed operation. The increased collector-substrate capacitance at the common collector node can cause circuit performance to degrade.

BiCMOS gate should swing from rail to rail to prevent excessive leakage. This can be done by adding either a PMOS device or NMOS device (or both) in parallel with the BiCMOS driver. An extra PMOS device will pull the output to the top supply rail while the NMOS will pull it to the bottom supply rail. By parallelling a CMOS inverter to the BiCMOS inverter, the output signal will go to both supply rails.

These devices, however, need to be small; otherwise excessive gate loading will be presented to the input of the BiCMOS gate and will degrade performance. The extra devices need only be large enough to pull the output to either rail within a reasonable time.

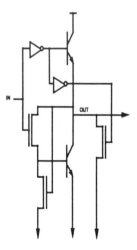

Figure 6.15 BiCMOS to CMOS level translator with NMOS pull down.

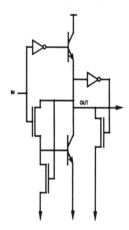

Figure 6.16 BiCMOS to CMOS level translation.

Fig. 6.17 shows a typical ECL OR/NOR gate used in decoding. Q2 is tied to a reference signal usually generated on chip. The output of the gate is driven by emitter followers. The output can be used to drive other ECL logic or it can be converted to a CMOS level signal in the manner discussed in the previous section. Care should be exercised in preventing the transistors from saturating. The output of the differential pair can be clamped with diodes to prevent saturation of Q1A, Q1B or Q2.

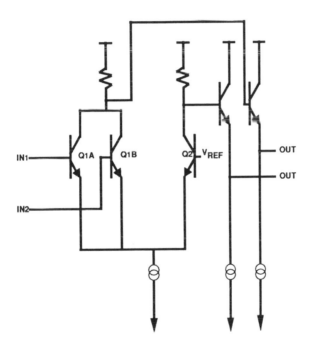

Figure 6.17 ECL NOR gate.

6.5.2 CMOS Decode

Basic gates used in CMOS decoding are NOR and NAND gates. In CMOS gates, an N input gate can be realized by stacking N devices in series and in parallel. In the actual utilization of CMOS gates in decoding (especially for high density chips), the number of inputs is limited. Too many devices in series reduces the effective drive current of the gate and thus slows down the rise time for NOR gates and the fall time for NAND gates. To compensate for the loss of drive current, the devices in series must be made wider. This increases the capacitive load presented to the previous stage as well as the area for layout. The larger size becomes critical in final decoders because of pitch constraints. Decoding gates must fit into pitches determined by the size of the memory cell. Also the increased load to the previous stage degrades performance. For 256 decoders using 3-input gates, the previous stage driver must drive either 32 or 64 decode gates. Therefore the increase in device size of the decode gate presents an increase in gate capacitance, multiplied by either 32 or 64, to the previous stage.

Sometimes a static pullup section is used in NOR gates to obtain a more symmetric transition and also reduce the loading on the previous stage. Fig. 6.18 shows a static PMOS load NOR gate. The gate of the single PMOS device in the pullup section is tied to VSS and the gates of the parallel NMOS devices are the inputs. A single PMOS device charges the load faster than a series of PMOS devices. Since the inputs are tied only to one device, the gate capacitive loading is essentially cut in half. In some applications [15] the extra power dissipated by the static PMOS load is more than compensated for by the decreased gate capacitance that has to be driven when operating the SRAM at high speeds.

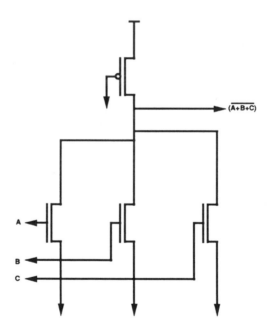

Figure 6.18 NOR gate with static PMOS load.

Often decoding is done in CMOS and then followed by a BiCMOS driver. An alternative is to use a BiCMOS gate to decode and drive simultaneously. This certainly is feasible, but for gates with three or more inputs there is an area penalty, which is quite costly when the gate has to fit in a certain pitch. The next section explains the reason for the area penalty when using BiCMOS logic gates.

6.5.3 BICMOS Decode

Any decoding implemented in CMOS can also be implemented in BiCMOS since the signal levels are basically the same. The difference between the two is drive capability and layout area. For driving large loads, BiCMOS decoding has an advantage over straight CMOS decoding. In addition to having greater driving capability, it also presents a smaller load to the previous stage.

Figure 6.19 shows a BiCMOS NAND gate. Note that the pullup section is a CMOS gate driving the pullup bipolar device in both cases. The pulldown section consists of series NMOS devices for the NAND gate. A parallel combination of NMOS devices in the pulldown section would be used for the NOR gate. Combining the decoder and driver into one gate enhances the performance, but in some instances is a tradeoff with chip area.

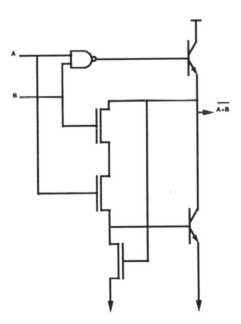

Figure 6.19 BiCMOS NAND function.

For three inputs, the combination of a CMOS gate and BiCMOS driver has 10 MOS and 2 bipolar devices while the BiCMOS gate also has 10 MOS and 2 bipolar devices. For four or more inputs, the number of devices in the BiCMOS gate exceeds that for the CMOS gate/BiCMOS driver combination. Obviously the reason for this is that all of the CMOS transistors are replicated in the pullup of the BiCMOS gate and all of the NMOS transistors are replicated a second time in the pulldown section. The number of devices per gate is of particular concern in the decoders of a memory since generally these gates must be layed out to match the pitch of the memory cell.

6.6 Memory Cells

In this section a number of memory cells are discussed. In the following we discuss the bipolar memory cell which has the advantage of high speed and the MOS memory cell which has the advantage of relative stability and margin. Also discussed is the use of the merged bipolar/MOS memory cell which attempts to capture the advantages of both of the above in a single cell.

6.6.1 Bipolar Memory Cell

The bipolar memory cell is based on the simple resistor-transistor inverter. This inverter only requires an input swing of 0.7 volts to cause the output to swing. Fig. 6.20 shows the basic bipolar latch which consists of two inverters connected back to back. The holding current for this cell must be set such that the NPN is

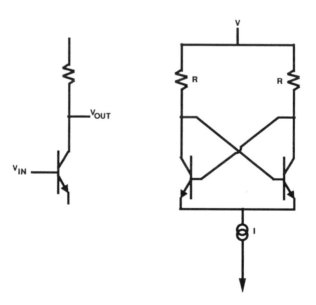

Figure 6.20 The basic bipolar latch

prevented from going into heavy saturation. At the same time the current must be large enough to maintain a suitable voltage drop across the resistor when the transistor is on. In fact the voltage swing is given as

$$IR_c = V_{off} - V_{on}$$

In order to read this latch an extra emitter is added to each transistor as shown in Fig. 6.21. When the memory cell is not selected, the potential of the bit lines is kept higher than the holding emitter potential. Thus no current flows to either bit line. The cell is selected by lowering the bit line voltage so that one of the emitters connected to the bit line becomes forward biased. This bit line current serves for the detection of the memory cell state.

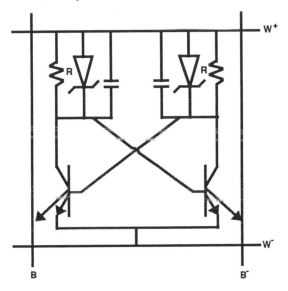

Figure 6.21 Basic bipolar memory cell.

In a large memory, the current per memory cell should be kept as small as possible. However, in order to maximize performance the read current flowing through the bit line emitter should be as high as possible. This conflicting requirement is resolved by the Schottky Clamp shown in Fig. 6.21. In the unselected state, the holding current is set just high enough to clamp the Schottky device. In the selected state, the higher required current is provided by the Schottky device. Thus forward biasing of the collector to base junction of the memory cell transistors is avoided.

Selection of the memory cell involves both the word line and the bit line. In order to read a particular memory cell, the upper word line is pulled high. Since the lower word line is just a current source, the lower word line potential will track the movement of the upper word line potential. The column select is performed by lowering the bit lines and thus forward biasing one of the two bases to bitline emitter junctions in the selected cell. The emitters on the bit lines which are part of the unselected cells do not conduct since their lower word line potentials are low enough to force conduction through the holding emitter.

Write selection of the memory cell differs from the read in that only one of the bit lines is pulled low. The selected bit line is pulled low enough to force one of the emitters to be forward biased. Once the bit line emitter is forward biased, the transistor will remain forward biased as the memory cell goes into the low holding current, standby mode.

The advantage of the bipolar memory cell is that the low voltage swings involved coupled with the relatively high read currents results in a very high performance SRAM. The disadvantage is the complexity of the control signal levels. In typical applications the voltage swing of the storage node is less than 500 millivolts and the total swing of the bit line is also less than 500 millivolts. This small voltage swing results in a relatively small value of charge stored on the internal nodes which is a concern from a soft error rate point of view. The introduction of high capacitance dielectrics such as Ta_2O_5 into the memory cell is an example of an innovation which addresses this issue.[6.16]

In order to gain even higher performance and improve stability, a coupling capacitor is often added to the bipolar memory cell. This coupling capacitor enables the internal nodes to follow the swing of the upper word line and hence improve speed. At the same the overall increase in the cell capacitance increases the storage charge and thus affords the cell greater stability. For some SRAM's the parallel combination of the Schottky clamp and resistor is replaced by a PNP load device. Major advantages of this cell are the lower array current and the typically smaller smaller size.

6.6.2 CMOS Six Transistor Memory Cell

The CMOS memory cell is based on the simple CMOS inverter. As shown in Fig. 6.22 two inverters are connected back to back with two pass transistors added to provide access. This cell is relatively stable since the difference between the logical level high storage voltage and the logic level low storage voltage is equal to the supply voltage. The word line is connected to the gate of the pass transistors and the bit line is connected to the drain/source diffusion of the pass transistors.

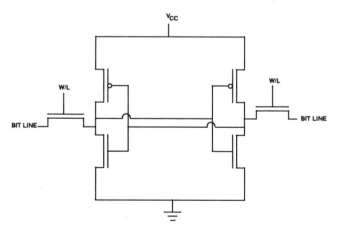

Figure 6.22 The CMOS memory cell.

When the six transistor CMOS cell is read, the bit lines shown in Fig.6.22 are held high. The cell is selected by the word line turning on the gate of the pass transistor. Current then flows from the bit line and through the driver transistor which is turned on. The conducting p-channel transistor does not impact the bit line read current. The read current is determined by the series combination of the pass transistor and the active driver transistor. Unlike the bipolar cell, a high read current is produced on every column that is associated with the word line. This can become a substantial contribution to overall SRAM power dissipation. In most large MOS SRAM's, care is taken to segment the word line so that excessive power dissipation is avoided.

During a write cycle one of the bit lines is pulled low. The low bit line forces that side of the cell to be pulled low. In order for the cell to trip (change state), it is necessary for the pass transistor is overcome the p-channel load device. Once the trip point is reached, the high internal gain of the cell will cause the internal nodes to clamp to the opposite supply rail. The bit line voltage at which this happens is called the trip voltage. In practice the trip voltage must be low enough to avoid unintentional writes during normal read operation.

The CMOS cell differs from the bipolar cell since in the standby mode no current flows and voltage switching rather than current switching is being utilized. In addition, during a write, both the word line and the bit line are required to swing by almost the full applied supply voltage. Thus this type of cell tends to be slower than the bipolar cell due to the high control voltage swings required. The high internal storage voltage yields a high storage charge for even moderate values of internal node capacitance.

6.6.3 Four Transistor, Two Resistor Memory Cell

The four transistor, two resistor memory cell is a special case of the CMOS six transistor cell. The p-channel pull-up device is replaced by a resistor as shown in Fig. 6.23. The purpose of the resistor is to provide the necessary current to keep the high side of the cell from decaying due to diode and subthreshold leakage. Typical values of resistance are in the Giga ohm range. This is usually sufficient to offset any parasitic leakage currents.

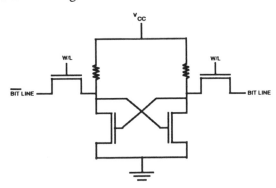

Figure 6.23 Four transistor, two resistor NMOS memory cell.

One of the important aspects of the four transistor, two resistor cell is the storage voltage. Due to the high resistance, the resistor can not be depended on to pull the high side of the cell to V_{cc}. During a write cycle the potential on the high side of the memory cell is set from the bit line through the pass transistor. As a result the nominal storage voltage is given by

$$V_{word} - V_T = V_{storage}$$

where the threshold voltage V_T itself is a function of the storage voltage as a result of the body effect. Since the storage voltage is reduced below V_{cc} the read current of the four transistor, two resistor cell is also reduced. This is because the storage voltage is the gate voltage of the driver transistor of the other inverter in the cross coupled node. This aspect of the four transistor cell is particularly important for BiCMOS since if a BiCMOS driver is used to drive the word line then the storage voltage will be further reduced due to the BiCMOS output node being a diode drop below V_{cc}.

The overriding advantage of the four transistor, two resistor cell is the reduced layout area. Normally the resistors are folded on top of the n channel devices and thus about a factor of two savings in layout area is achieved.

6.6.4 Merged Bipolar/CMOS Memory Cell

The merged bipolar/CMOS memory cell attempts to combine the low standby power of the six transistor CMOS cell with the high read current of the bipolar cell. Such a cell is shown in Fig. 6.24. If node 2 is held high then the read current coming from the p-channel device is amplified by the current gain of Q6. Selection of the cell is provided by the read word line going high and the read bit line going low. Since the logic swing is small, speeds comparable to that of the bipolar memories can be achieved. Single ended reading is accomplished by comparing the internal voltage level of the selected cell with that of an on-chip reference voltage. This reference voltage is generated in such a way that it tracks the voltage level of the selected read word line.

In terms of the write cycle, it is necessary for the logic level of both the write word line and the write bit line to swing by the full supply voltage. Like the read cycle, the write cycle as designed in [6.1] is also single ended. Thus unlike the conventional CMOS memory cell, the pass transistor is required to pull the internal node either high or low. Due to the need for the use of CMOS logic levels in the write cycle, the write cycle is slower than the read cycle. However, the high speed read cycle and the ability to do simultaneously read and write operations makes this cell attractive for some applications.

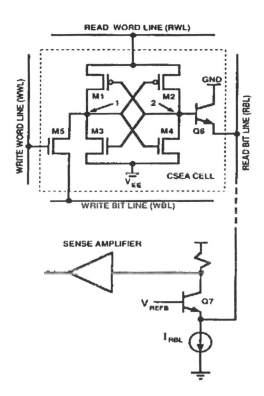

Figure 6.24 Merged memory cell after Yang et al. [6.1].Reprinted with permission of the publisher. (© 1988 IEEE)

6.7 Sensing

6.7.1 Standard Sensing Operation

A typical sensing scheme for MOS RAM's is show in Fig. 6.25. As seen from Fig. 6.25 the read current develops a differential signal on the bit lines. The NMOS source follower serves to provide the load for the cell read current. Since the read current flows to only one side of the memory cell, a differential voltage on the bit lines is developed. The pass transistors M4 and M5 allow the signal to reach the main sense amplifier.

In the general case the M4 and M5 pass transistors provide the multiplexing for the selected column. Thus many columns will share the same main sense amplifier with the pass transistor serving to switch in the selected column. At the 256K SRAM level of integration typically 32 columns will share a single sense amplifier. In this scheme it is necessary for the memory cell to drive the bit line and in addition drive the drain junctions of all the unselected columns' pass transistors. The most standard implementation of BiCMOS is to replace the MOS main sense amplifier with a bipolar sense amplifier. The bipolar sense amplifier, because of its

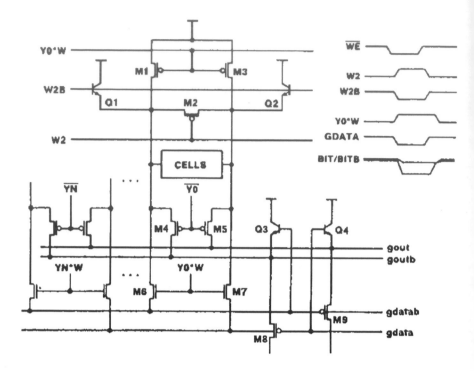

Figure 6.25 Typical BiCMOS sensing scheme after Kertis et al.[6.11]. Reprinted with permission of the publisher. (© 1988 IEEE)

superior differential amplification capability, lessens the bit line voltage swing necessary to develop an output signal. Since cell current remains the same, the overall SRAM performance is improved. For ECL SRAM's this is particularly attractive since the amplifier output can be made ECL compatible.

6.7.2 Individual Column Amplification

An advancement recently utilized for BiCMOS, is the use of a bipolar sense amplifier for each individual column. This requires a high packing density for the individual transistor elements. Since this sense amplifier must fit within the pitch of the memory cell implementation of this scheme is a major challenge in circuit layout. The circuit schematic for such a scheme is shown in Fig. 6.26.

During the read cycle, current is pulled from the pull-up bipolar transistors and flows down into the memory cell. The resistor and the change in the base-emitter potential serves to provide the bit line differential signal. Since both the DATAIN and $\overline{\text{DATAIN}}$ signals are low, the bit line differential voltage appears at the respective bases of the bipolar differential pair. Collector current flows in the bipolar device which has the higher base current.

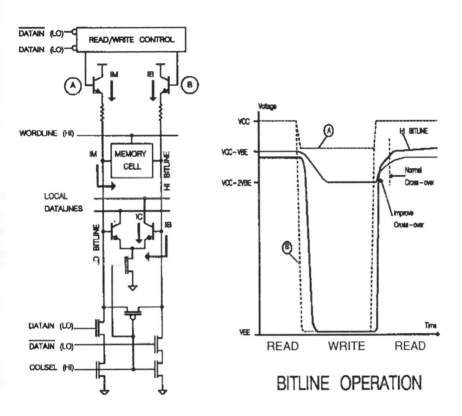

Figure 6.26 Individual column amplification sensing scheme.

Although each column has its own differential pair, current only flows in the differential pair of the selected column. The current source of the bipolar differential pair is provided by a NMOS device. This NMOS device is gated by the column select signal. When the column is not selected, both bit lines are shorted together by a p-channel device. Each bipolar device in the differential pair has its collector connected to one of the local data lines. Thus all columns in a 32 column block have all their differential bipolar devices connected in a two common collector configuration. Current only flows in the bipolar device which has its column selected and has its base connected to the higher side of the bit line differential voltage.

The individual column sensing scheme is inherently high performance since the memory cell is only required to drive the capacitance of the bit line to achieve the necessary sensing of signal. The parasitic capacitance normally associated with multiplexing the columns to a main sense amplifier is now driven by the high performance bipolar devices. This design technique segments the capacitive loading at each stage of amplification such that the capacitive loading is appropriate to the device driving the load.

Writing of data into an individual bit is accomplished by forcing the bit line low with the use of the series combination of two n channel transistors. This is illustrated in Fig 6.26. Both the $\overline{\text{DATAIN}}$ and the COLSEL signals are high. At the same time the bipolar pull-up transistor is turned off. Thus the full bit line capacitance is discharged through the two serially connected NMOS transistors. With the pass transistor of the selected cell turned on, the low bit line voltage will cause the memory cell to trip.

After the write is completed, the low bit line is recharged back to its original potential by turning on the bipolar pull-up device. The use of the bipolar device results in a fast recovery back to the read state. This fast write recovery is further aided by slightly dropping the high bit line during the write cycle so that both bit lines rise together. The use of a bipolar differential sense amplifier ensures that any common mode noise is ignored and thus reading of valid data occurs sooner.

6.7.3 DRAM's Amplification Scheme

Direct sensing of the small signal from the bit line of a DRAMs memory array by coupling the bit line to the base of bipolar transistor, has not been employed to date. This is because of the base current required to turn on the bipolar transistor and the difficulty in designing the complex circuitry to restore the bit line after the read cycle. Unlike the SRAM memory cell, which provides a large current differential on the bit lines, the DRAMs charge storage nature requires the sensing scheme to differentiate the difference of small charges on the bit lines. An isolator circuit is preferred to separate the high impedance bit line from the current hogging bipolar transistors.

One of the approaches to DRAMs bit line sensing is shown in Fig. 6.27 In this design, the MOS preamplifiers is placed between bit lines and the bipolar differential sense amplifiers to provide both electrical isolation and signal amplification[6.5]. The bipolar sense amplifier serves as an intermediate amplifier and signal driver. The main advantage from this approach is that the signal from the bipolar sense amp can be directly used in the next stage logic operation. In CMOS DRAMs the latching sense amps require another psuedo SRAM type like sensing stage before output signals can be used in later stage logic operations. ECL I/O DRAMs will also have an advantage in read cycle access time, which is a benefit from avoiding conversion of the sensing signals to full CMOS levels.

Other considerations for DRAM bipolar sense amplifiers include the silicon area needed for the sensing scheme layout. The minimum bipolar transistor size is much larger than that of a minimum size MOS transistor. Bipolar transistor isolation spaces become much more important in the pitch layout circuitry than do MOS transistors. This is an area ready for significant innovation.

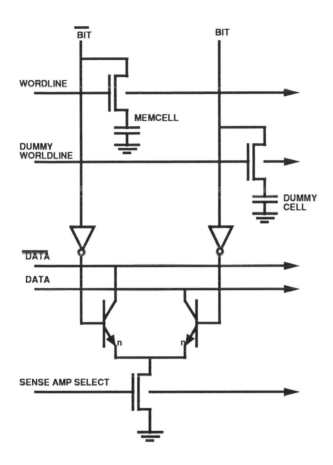

Figure 6.27 DRAM sense amplifier.

6.7.4 Final Sense Amplification

There are two major types of circuits used in the final sense amplification for BICMOS ECL I/O SRAMs using MOS memory cells. The dual cascode circuit (Fig. 6.28) used in many high density SRAM designs offers fast access times by reducing the voltage swing of the read data line[6.10]. The signal from first sense amplifier is collected on this read data line which results in a high capacitance load. Reducing the voltage swing of this line has a direct effect on the signal delay. However, because the amplification action is at the collector of the cascode transistor, any noise coupling along the data line will also be amplified along with the signal. Speed is traded off for higher signal margin in the second final sense amp design shown in Fig. 6.29. Common mode noise is filtered out and not amplified as in the first design. This sense amp uses diodes to reduce the data line voltage swing to the VBE differences between transistors Q1 and Q2 [6.11].

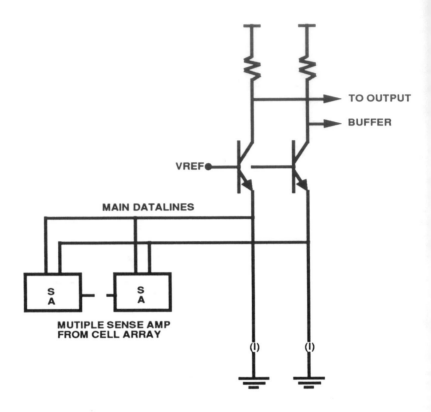

Figure 6.28 Cascode sense amplifier.

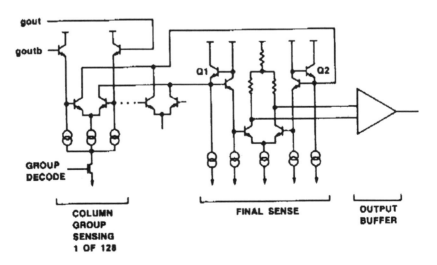

Figure 6.29 Low voltage swing final sense amplifier as proposed by Kertis et al. [6.11]. Reprinted with permission of the publisher. (©1988 IEEE)

For the TTL I/O BiCMOS SRAM, speed is greatly enhanced if the small signal swing from the first sense amp is retained through to the last stage before converting to full CMOS levels for output. The final sense amp can be designed to be one circuit which collects signals from previous sense amps, amplifies these signals and then converts them to a CMOS level to drive the output pad.

6.8 Output Buffer

The output buffer is one of the circuits which has received much attention in SRAM's designed for both ECL and TTL output levels. Noise and ringing can more than just degrade performance in many products. Improper output design can cause the RAM to fail to meet specification and in some cases cause catastrophic failures. Moreover, shortcomings in the output buffer design may not be readily apparent until the circuit is placed in the system environment. BiCMOS technology offers more solutions to the output buffer than CMOS technology. However, much more improvements are needed to accommodate the higher performance typical of new generation SRAM's.

6.8.1 ECL Output Buffer

The BiCMOS technology has the same advantage on the ECL output buffer as on the generic ECL input buffer. Furthermore, the output signals from the sense amplifier of a memory chip often are small voltage signal swings. They can be directly transferred to a compatible ECL output signal without passing through any large voltage swing amplification stages as in a pure CMOS memory design. Standard ECL output buffer such as one shown in Fig. 6.30 are commonly used in the output stage of the ECL I/O BiCMOS memory chip.

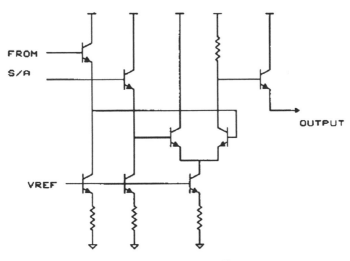

Figure 6.30 ECL output buffer.

For BiCMOS memory devices which have a full CMOS signal available at the output stage, a simple CMOS to ECL converter discussed in 6.3.2.2 can be used to provide an ECL output signal.

Many attempts have been made to use the MOS transistor for producing ECL output levels to drive a 50 ohm transmission line load [6.8]. The lower g_m problem of the MOS transistor has been partially solved by using an oversized MOSFET, or a MOSFET with an extremely short length, to produce the current required. In addition, multiple power supply voltages are employed to provide the gate voltage required for the output transistor. However, the sensitivity of the output levels to temperature remains an unsolved problem and will require more complex and non-standard solutions. Ironically, these solutions may increase the system cost of the CMOS design many fold over the initial increase in process cost in going from a CMOS process to a BiCMOS process.

6.8.2 TTL Output Buffer

Replacing the MOS pull-up device in the output stage of the TTL output buffer with the bipolar transistor is a common practice in BiCMOS TTL output buffer design. It is much faster to pull up a high capacitance load with the bipolar transistor which is working in the emitter follower configuration than with the MOS transistors. However, for pulling down the load to the TTL I/O level, a MOSFET serves as well as an NPN.

More and more system users require the current sourcing of the output transistor to meet the specification of up to 25ma at 0.4 volt TTL low level. The bipolar transistor can be used to provide much greater current sourcing than the MOS transistor at much smaller layout area and lower loading for the previous stage driver. However, at output levels below 0.4V there is the danger of the output bipolar transistor going into saturation. This not only degrades the performance but also could cause a catastrophic latch-up failure. At the cost of process complexity, Schottky diodes are often used in the output stage to prevent the base collector junction of the output transistor from becoming heavily forward biased.

Fig. 6.31 provide another solution for the problem with the cost of design complexity and transient operation uncertainty through the use of feedback [6.17]. The DC low level of the output (V_{outL}) is clamped at

$$V_{outL} = V_{BE} \frac{R\,1}{R\,2}$$

When the output is above V_{outL}, most of the current flow is to the base of the pull-down output transistor which provides the required sink current for the output load. Node A is clamped at $2V_{BE}$ by the diode D1 and V_{BE} of the output transistor.

However, when the output terminal falls below VoutL, the current flowing through the feedback network will increase until the the current flow to the base of the pull-

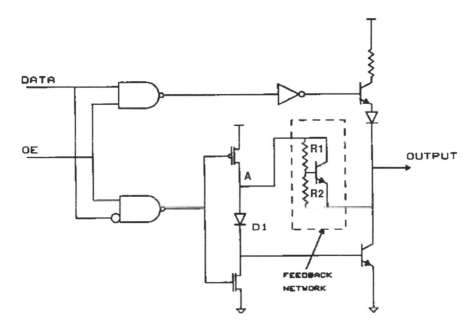

Figure 6.31 BiCMOS TTL output with feedback network.

down transistor is reduced to the point that the output terminal voltage level stops falling any further. It can be seen that the low output voltage level will stop at

$$V_{outL} = V_A - V_{fb}$$

where V_A is the voltage at node A and V_{fb} is voltage set up by the feed back network.

6.8.3 CMOS Output Buffer

The CMOS I/O signal which has the maximum signal margin also derives much benefit from the BiCMOS technology. The P and N channel transistors at the output stage place a large capacitance load on the previous stage. Also logic operations are often required at the output stage in the memory chip for separate read and write. BiCMOS gates provide drive capability in combination with logic operations and result in fewer stages than the counterpart CMOS design. Fig. 6.32 shows the commonly used BiCMOS output buffer which provides full CMOS level or TTL compatible output levels for interfacing with both CMOS and TTL I/O signals.

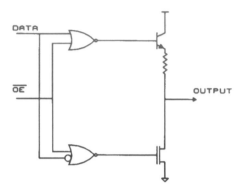

Figure 6.32 BiCMOS TTL output buffer.

6.9 General Noise Considerations

Combining analog type ECL circuits with digital type BiCMOS/CMOS logic circuits on one silicon chip requires careful design and layout techniques to ensure that the noise generated from the digital circuitry will not affect the operation of the analog type circuitry. Reference voltage and current source circuits are good examples of circuits which can be very sensitive to on-chip noise. These types of circuits are found in abundance in ECL circuit blocks. Many techniques to address this problem already exist in the design of standard CMOS mixed analog/digital chips. They include the isolation of analog from digital power supplies, physical layout techniques using guard rings to minimize the noise propagated through the substrate and circuit design considerations to avoid direct capacitive coupling and other charge injection effects. One example of such a circuit is the Vbump circuit in a BiCMOS SRAM [6.12]   This circuit uses an analog type filter circuit to minimize the effect of VCC noise on the bit line. Care also should be taken to avoid direct coupling from wide voltage swing signal lines to analog signal lines. Shielding layout techniques which have been widely practiced in PC board layout design could be applied to the critical signal lines in a high speed mixed ECL/ BiCMOS chip to improve the noise imunity.

6.10 Summary and Discussion of BiCMOS Implementation

A typical delay path of a row address access for a 256K ECL SRAM is shown in Fig. 6.33 [6.12]. In this delay path, CMOS logic levels are involved only in the global word line drivers and decoders as well as the local word line drivers and decoders. The rest of the delay path involves propagation of signals using ECL levels or delays through the level translator. Thus although the majority of the SRAM's control signals use CMOS logic levels, most of the critical delay path is limited by the speed of ECL circuits which are very bipolar intensive. Typically we have found that circuits which operate using CMOS logic levels are very sensitive to variations in MOS device parameters while circuits involving ECL logic levels are sensitive to variations in bipolar device parameters.

Of note in Fig. 6.33 is the relatively large amount of time consumed in ECL to CMOS level translation. The choice of when in the delay path to do the ECL to CMOS level translation is set by power dissipation considerations. For this particular application, there are 256 rows and only one of these is selected at any given time. Thus the use of high standby current ECL circuits for the global word line drivers and decoders is precluded due to the low duty cycle. However, the predecoders were chosen to operate using ECL circuitry because these circuits are fewer in number and are active a much larger percentage of the time.

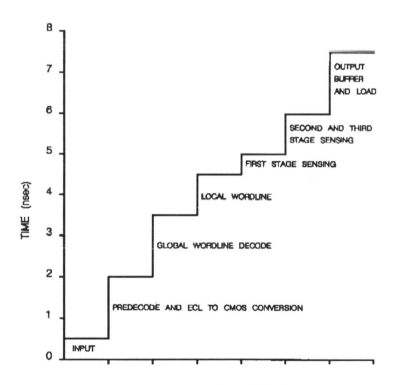

Figure 6.33 Critical delay path for a 256K SRAM.

After the state of the selected memory cell is sensed, the signal propagates to the output using differential ECL circuits. However, the associated control logic along this signal path has been done using CMOS logic levels. This combination allows for the use of ECL signals while being able to power down unused circuit blocks. Thus in a given read operation only the amplifier circuits involved in signal amplification for the specific address being accessed are active.

6.11 Outlook

The application of BiCMOS to SRAM's has been successfully demonstrated by many companies. However, it remains to be seen if BiCMOS can make an impact on the DRAM's. This is mainly because of the cost and density issues. DRAM's can make use of the drive capabilities, ECL interface, and process/temperature insensitivity that BiCMOS technology offers. As long as the DRAM's speed is limited by the multiplexed address scheme, the speed improvements to be made by using BiCMOS will not be significant enough to warrant extra cost in a commodity market. Some niche markets will, on the other hand, use BiCMOS technology to their advantage. Additionally, as the design community gains more experience with the BiCMOS technology, new circuit techniques will flourish. The BiCMOS SRAM design will move more aggressively towards clocked designs taking advantage of the performance stability against the process and temperature variations. The use of BiCMOS will migrate to the non-volatile memories as well.

References:

[6.1] T.S.Yang et al.,"A 4-ns 4K x 1-bit Two-Port BiCMOS RAM,"JSSC Vol.23, pp. 1030-1040, Oct. 1988

[6.2] B. A. Chappell et al., " Fast CMOS ECL Receivers With 100mV Worst-Case Sensitivity," IEEE J. Solid State Circuits, vol. 23, pp. 59-67, Feb. 1988.

[6.3] M. Yoshimoto et al.," A 64Kb Full CMOS RAM with Divided Word Line structure", Digest of Tech. Papers. pg.58 , ISSCC, Feb. 1983 N.Y

[6.4] R. Hori et al.,"An experimental 35ns 1Mb BiCMOS DRAM's", Digest Of Tech. Papers. p. 280, ISSCC. Feb. 1987, N.Y

[6.5] S. Watanabe et al., "BiCMOS Circuit Technology for High Speed DRAM's." 1987 Symposium on VLSI Circuits, pp. 79-80, Aug. 1987, Karuizawa, Japan

[6.6] E. Hudson and S. Smith, "An ECL compatible 4K CMOS RAM," in ISSCC Dig. Tech. Papers, vol. XXV, Feb. 1982, pp. 248-249.

[6.7] Y. Ohmori et al., "An ECL compatible 64Kb FIFOS/CMOS static RAM," in Extended Abst. 17th Conf. Solid State Devices and Materials (Tokyo), 1985,   pp. 53-56.

[6.8] T. I. Chappell et al., " A 6.2ns 64Kb CMOS RAM with ECL Interfaces,"   Symp. on VLSI Circuits Dig of Tech. Papers, Aug. 1988, pp. 19-20.

[6.9] S. Miyaoka et al., " A 7ns/350mW 64K ECL Compatible RAM," in ISSCC Dig. Tech. Papers, vol. XXX, Feb. 1987, pp. 132-133.

[6.10] K. Ogiue et al., " A 13ns/500mW 64Kb ECL RAM," in ISSCC Dig. Tech. Papers, vol. XXIX, Feb. 1986, pp. 212-213.

[6.11] R. A. Kertis et al., " A 12ns 256K BiCMOS SRAM," in ISSCC Dig. Tech. Papers, vol. XXXI, Feb. 1988, pp. 186-187.

[6.12] H. V. Tran et al., " An 8ns Battery Back-up Submicron BiCMOS 256K ECL SRAM," in ISSCC Dig. Tech. Papers, vol. XXXI, Feb. 1988, pp. 188-189.

[6.13] H. V. Tran et al., " A Novel BiCMOS TTL Input Buffer; A Merging of Analog and Digital Circuit Design Techniques," Symp. on VLSI Circuits Dig of Tech. Papers, Aug. 1988, pp. 65-66.

[6.14] Y. Sugimoto et al., "Bi-MOS Interface Ciruit in Mixed CMOS/TTL and ECL use Enviroment," 1st. Int'l. Symposium on BiCMOS Technology and Circuits, May 1987 (ECS).

[6.15] K. Sasaki et al., "A 15 ns 1Mb CMOS SRAM," in ISSCC Dig. Tech. Papers, vol. XXXI, Feb. 1988, pp. 174-175.

[6.16] N. Homma et al., "A 3.5-ns, 2-W, 20-mm2, 16-kbit ECL Bipolar RAM," IEEE J. Solid State Circuits, vol. 21, pp. 675-680. Oct. 1986.

[6.17] M. Nakashiba et al., "A Subnanosecond BiCMOS Gate-Array Family," Proceedings of IEEE 1986 CICC, pp. 63-66, May 1986

Chapter 7

Specialty Memories

C. Hochstedler
(National Semiconductor)

7.0 Introduction

Specialty memories offer features and functions beyond those found on generic memory. It is these special features and functions which bring benefit to particular applications, and often also render the specialty memory virtually useless in other applications. The applications for specialty memory are widely varied, and make the study of specialty memory quite interesting. Application specific memory would be a good descriptive name, except ASIC memory, through common usage, usually implies custom memory implemented within a standard cell or gate array ASIC device.

This chapter will define specialty memories, and discuss several specific types of specialty memories. While several types of specialty memories will be discussed, the intent is not to provide a complete list of all specialty devices available or conceivable. A complete list is quite impractical. The intent is to provide a good understanding of the breadth and diversity of specialty memories. Within this scope, the benefits afforded by BiCMOS process technology for specialty memory devices will be addressed.

7.0.1 Definition Of Specialty Memories

Specialty memories are not easily defined. It is much easier to define what types of memory devices are not specialty memory. Examples of specialty memory are, however, easy to list. Prior to defining specialty memories, an explanation of standard memories and some examples of both standard and specialty memories will be used to bring the issues into better focus.

Standard memories could be defined as follows:

Standard memories are DRAM or SRAM memory devices with only the following narrow set of functions and features:

Generic DRAM

Volatile (requiring power and refreshing)
Synchronous Interface
Multiplexed Addresses
Separate Data In and Output
RE (RAS), Row Enable (Row Address Strobe)
CE (CAS), Column Enable (Column Address Strobe)
W, Write Enable

Generic SRAM

Volatile (requiring power)
Asynchronous Interface
Non-multiplexed Addresses
Separate or Common Data Input and Output
E or S, Enable or Select
W, Write Enable
G, Output Enable (on some generic SRAM's)

This explicit definition of standard memories helps provide an understanding of specialty memories. The distinction of a standard memory is that it conforms to a fairly narrow industry standard set of features and functions, rather than the type of memory cell. This definition is also notably devoid of distinction according to device organization (e.g. 1M x 1, 256K x 4, 128K x 8 organizations of 1M density devices). It also makes no distinction regarding performance, in terms of speed or power. Package options and I/O levels (i.e.. TTL I/O, ECL I/O) are also excluded from the definition.

The generic industry standard is not a constant set of features and functions, forever unchanging. Many early SRAM's did not offer the output enable control, but today it is so common that it is included in the list of generic features. Multiplexed addresses on DRAM's were not introduced until the 16K density generation, but quickly became the standard. The generic industry standard definition evolves as the industry progresses.

ECL SRAM's, for example, can be standard memories under this definition. Many ECL SRAM's offer only the industry standard set of functions. Some ECL SRAM's contain unique and special features and functions, such as STSRAM's (Self - Timed SRAM's). ECL STSRAM's fall well outside the range of this definition, and must be considered specialty memories. This example points out that the distinction is based on only the features and functions. Video DRAM, or VRAM, is another good example. VRAM's contain specialized circuit functions not found on the generic DRAM, such as a high speed serial shift register and special reading and writing modes. These special features and functions are included for the benefit of the video display system designer. Because these features provide little benefit in most other applications, it is proper to classify the VRAM as a specialty memory.

It makes sense to consider specialty memories as application specific memories. In this context application specific is intended to describe the usefulness within a very narrow range of applications. Multiport register file memories are an example. A 64 x 16 register file memory featuring two write ports and three read ports is certainly a specialty memory; very well suited for register file applications, but very poorly suited for generic main memory. In general, a test for specialty memories could be to consider whether the device is best described as application specific, or as generic.

A mask programmed ROM is programmed permanently with a given data pattern during wafer fabrication, usually with a unique contact or gate mask for each data pattern. The ROM, considered as a function block, is most often generic in features and functionality. The data pattern which the ROM contains is certainly unique and application specific. A ROM can be easily patterned to be used as a code converter; a binary to BCD code converter pattern is a simple (if not very practical) example. When programmed and applied as a code converter is a ROM still a standard memory? Or is it better described as a memory with the features and functions of an ASIC logic block? In the case of the volatile memories previously defined (SRAM and DRAM) there was no consideration of the data pattern, or the use of any one particular data pattern. It is most logical to expand the definition of generic standard memory devices to include ROM and PROM (fuse PROM, UV EPROM, EEPROM, etc.) including only those devices with the generic set of features and functions common for that type of memory.

If standard memory is any common type of memory which contains only the industry standard features and functions, then specialty memory must be all other types of memory. Some examples of specialty memory may help to illustrate the broad scope of the classification:

Examples of Specialty Memories

Video DRAM
Latched and Registered SRAM
Self - Timed SRAM
Content Addressable Memory
Multiport SRAM
FIFO

Defining specialty memories as nonstandard memories is not proper because it really does not delineate the group. Essentially the same, but couched in better wording to clarify the basis of the distinction is the following definition:

> Specialty memories are a broad spectrum of memory devices embodying features and functions which differentiate them from standard memories, giving them added value for specific applications, or for relatively narrow ranges of applications.

7.0.2 Why Specialty Memories Exist

Specialty memories are forecast to grow relative to standard memories over the coming years [7.1]. The primary reason for the growth is system performance (speed) enhancement, and secondarily, cost reduction. A FIFO, for example, can be implemented with software and standard DRAM. This implementation is relatively slow. It can also be done with logic gates (i.e. counters, multiplexers, etc.) and standard SRAM. This requires several packages, higher power, more circuit board space, and so on. A monolithic FIFO memory offers a single device solution with high speed performance at lower cost than most alternatives. Specialty memories offer benefits of higher performance in specific applications than standard memories, and integration of logic functions on the memory device.

BiCMOS is an important technology for specialty memories. The predominant driving force for specialty memories is enhanced speed performance in specific applications. BiCMOS offers the memory designer more latitude to optimize for speed than an equivalent, or even more advanced CMOS process (e.g. BiCMOS provides a bipolar device which can switch a highly capacitive node more quickly than a MOSFET). BiCMOS also allows the high performance memory device designer significant latitude in integrating logic elements on the memory die in CMOS, which is usually much more compact and lower power than bipolar logic. I/O level flexibility (choices of ECL I/O, TTL I/O, CMOS rail-to-rail I/O), and speed versus power and density choices are good examples of the flexibility BiCMOS brings. In short, the added flexibility of BiCMOS allows the high performance memory circuit designer more freedom to tailor the device for specific application requirements.

7.1 Standard SRAM

A familiarity with standard SRAM features, functions and operating modes is a prerequisite for an exploration of the differences found in specialty SRAM. Chapter 6 has provided a good foundation for understanding the circuitry inside a standard SRAM. This section will briefly review the operation of the standard SRAM from an external (applications) viewpoint. The standard SRAM is a very straightforward device because it is asynchronous and has only a few control inputs. A representative generic SRAM is illustrated in block diagram form in Fig. 7.1.

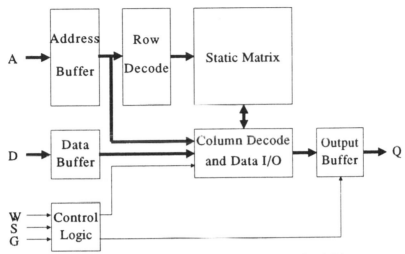

Figure 7.1: Standard Asynchronous SRAM Functional Block Diagram.

Normal control inputs are S, chip select, and W, write enable. Also shown is a third control, G, output enable. Output enable is present on many standard SRAM devices, but not all; it is considered optional within the definition of standard devices. (Appendix A is provided for the convenience of readers not familiar with the JEDEC / IEEE conventions for pin and function nomenclature used in block diagrams and timing diagrams throughout this chapter.)

Key to understanding the operation of the asynchronous SRAM is the recognition that there are no latches or registers, and no associated clocks or latch enables (except of course the SRAM cell itself, which is a latch). An input signal transition propagates into or through the device and causes a logical effect completely irrespective of the state of any other input state. The timing diagrams in Fig. 7.2 clarify the asynchronous operating modes.

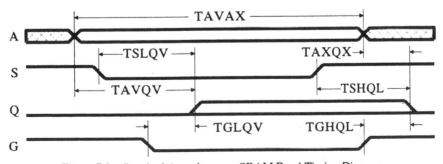

Figure 7.2a: Standard Asynchronous SRAM Read Timing Diagram.

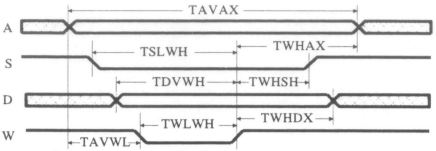

Figure 7.2b: Standard Asynchronous SRAM Write Timing Diagram.

Throughout this chapter ECL I/O style timing diagrams are shown because many of the applications for BiCMOS memories are found in ECL I/O systems. Many of the early BiCMOS memories developed were standard ECL I/O SRAMs [7.2]. There is little difference between a timing diagram for a TTL I/O device and an ECL I/O device. The differences are in the output waveforms; TTL I/O diagrams illustrate the high impedance state, while ECL I/O diagrams show an inactive output as a low level (VOL).

Asynchronous operation of the standard SRAM is illustrated by the read cycle. As long as the SRAM is enabled it will respond to any address change after one address access time delay. This implies that there is no latching or registering of the address inputs, and no latching or registering of the data outputs, by the control signal inputs. Some memory devices act as if they are asynchronous, and yet contain latches and internally generated timing controls (e.g. address transition detection). Recall, however, that the definition draws the distinction based on the features and functions offered to the system designer; not on the chosen internal circuit implementation.

7.2 Synchronous Specialty Memories

Synchronous memories are distinguished from asynchronous memories by the existence of a clock function input. The clock may not be explicitly called a clock, and may also be used to generate other control functions, but it still is a synchronizing control. A memory with an address register requires a control signal to set the timing when the register captures the address information. Even if this pin is called ALE (Address Latch Enable) or E (chip Enable) it performs the function of clocking the address register, and initiates the memory cycle. Asynchronous memories, in contrast, have no synchronizing or clocking signal input. They respond to a change in address inputs without any other change in control signals required. Commonly available synchronous memories include industry standard DRAM, Registered SRAM, Latched SRAM, Self - Timed SRAM and Registered PROM.

7.2.1 Latched and Registered Synchronous SRAM

Commonly available latched and registered synchronous SRAM differ from standard SRAM by the feature of on chip address latches or registers, and the necessary register clock or latch enable control input. There are several variations on the same basic theme. Some register addresses only; other devices latch address inputs and data inputs. A few also provide data output latches. A representative device functional block diagram is shown in Fig. 7.3 . The illustrated device features addresses registered by the falling edge of C, the clock input. The data output is latched when the clock is held high, and can be disabled by the output enable control, G.

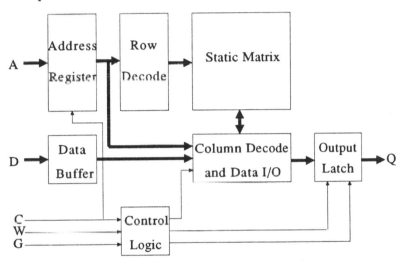

Figure 7.3: Registered Synchronous SRAM Functional Block Diagram.

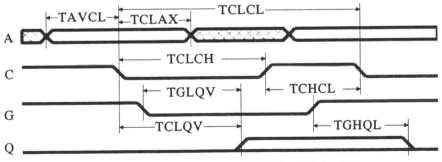

Figure 7.4a: Registered Synchronous SRAM Read Timing Diagram.

Important differences from the standard asynchronous SRAM are noticeable in the block diagram, but they really show up in the timing diagrams of Fig. 7.4. The difference in the address bus timing is readily apparent in a comparison of the timing diagrams (Fig. 7.2 and 7.4). The address register of the synchronous

SRAM captures the address at the time of the clock transition. The address bus is then free to change to the next address, in preparation for the next memory cycle. The flexibility to set up the next address earlier (than possible with an asynchronous SRAM) can be an advantage for some system designs.

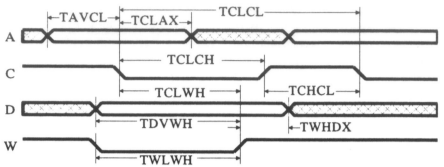

Figure 7.4b: Registered Synchronous SRAM Write Timing Diagram.

Comparison with the asynchronous timing diagram also reveals differences in the data output timing. The illustrated synchronous device offers a flow through type data output latch, which holds the outputs valid after the clock returns high. Cycle time can be reduced as a result of the data output latch, because the clock may return high as soon as read access time is met. System constraints often require data output to be valid for a significant period of time after access has occurred. Latching the outputs allows two timing constraints to be met simultaneously; the system data path constraints and the RAM minimum clock high time. This overlapping of the two timing constraints may reduce the cycle time in the system by approximately the magnitude of the lesser of these two constraints.

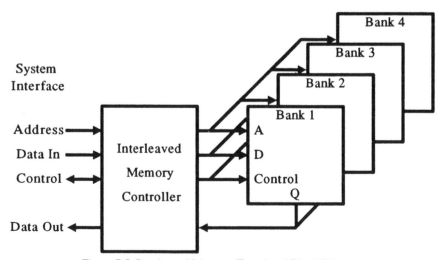

Figure 7.5: Interleaved Memory Functional Block Diagram.

Applications and Benefits. Interleaving is one technique commonly employed to improve memory bandwidth in very fast computing applications. Interleaving amounts to subdividing memory into banks designed such that memory accesses to several banks occur in an overlapping manner, slightly staggered in time. The theoretical cycle time of an interleaved memory bank is the cycle time of one bank divided by the number of banks. Because of the additional logic involved (e.g. multiplexers), the cycle time of a bank is usually slightly longer than the cycle time of a non-interleaved memory implemented with the same RAM device. For example, a 12 ns ECL I/O BiCMOS SRAM should be usable with a 20 ns cycle assuming the system is carefully designed to control signal skews. A two way interleaved design could achieve 15 ns cycle or better; falling short of the theoretical 10ns. A four way interleaved design should reach 8 to 10 ns cycle. Figure 7.5 illustrates a block diagram of a four way interleaved memory array.

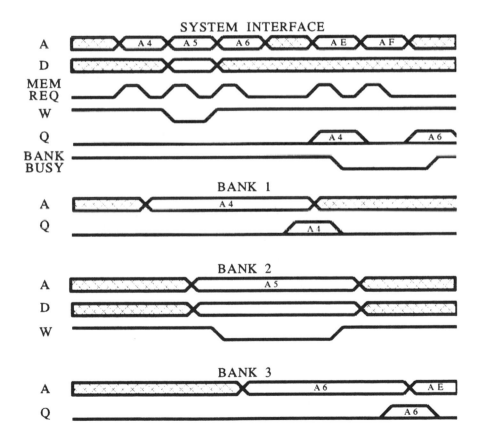

Figure 7.6: Interleaved Memory Simplified Timing Diagram.

Interleaved memory can be effective in increasing the memory bandwidth, but not without some compromises. If, for example, the next access references a location which is physically in an interleave bank which is already busy accessing, a delay is incurred. Assignment of the address space can influence the performance. If the memory field is decoded in such a way that the least significant address weights determine which of the interleaved banks is addresses, then highest bandwidth is achieved when accessing contiguous memory blocks (in address weight). In the four way interleaved example, the worst case will occur when the desired data are to be found at addresses exactly four addresses distant from each other. Because code and data are most often required sequentially the address assignments can be chosen to minimize the bank busy problem for sequential accesses. Figure 7.6 shows a simplified timing diagram of a sequence of memory references with a bank busy delay for bank 3. Bank 4 timing is not illustrated because it is not accessed for this example.

Notice that the address inputs to the interleaved array arrive at a rate which is faster than the cycle time of the RAM. Address latches or registers are required, in the case of standard SRAM, to hold the address stable for the cycle time of memory within a given bank, while the address bus from the processor makes several transitions to address each other bank in sequence. Synchronous RAM, because of the on chip latches or registers, can be beneficial in simplifying the design by eliminating the need for a bank of discrete registers, usually improving the speed. State of the art designs invariably result in custom ASIC gate arrays or standard cell based logic for most of the memory support logic needed to implement interleaving. Interleaved memory is not common for a broad spectrum of applications. The design complexity represents a price paid for the speed benefit, which relegates it to only speed critical applications. In the applications where interleaving is advisable, synchronous RAM is often the preferred choice.

There are several variations of synchronous memory. A few more specialized types are discussed in later sections. Latched or registered, address inputs only or address inputs and data outputs, or any other combinations; the driving force is invariably speed performance gained through system architecture innovation. BiCMOS becomes a strong candidate for the process of choice because of the demand for speed. Many of the fastest computing applications are ECL logic based machines, making BiCMOS an attractive choice for densities above the practical limits of bipolar ECL memory. Internally, synchronous memories may employ dynamic logic techniques for speed, which are much more readily implemented in MOS than in bipolar. For example, in a clocked SRAM one critical path is the on chip clock distribution because the clock usually is required to drive considerable capacitive load. For an ECL I/O device the clock input and all the address input registers could be implemented in bipolar ECL for high speed. Even for a TTL I/O RAM, bipolar clock drivers could be employed to speed up the clock distribution (as compared to CMOS). Speed and ECL I/O level compatibility, coupled with enhanced design flexibility in the on chip logic, combine to make BiCMOS a preferred choice for high speed synchronous RAM designs.

7.2.2 Self - Timed SRAM

Self - timed SRAM is a type of synchronous SRAM with an important additional distinguishing feature; on chip write pulse timing circuitry. Most self - timed memories register or latch every input signal (addresses, data, and control signals) with an input clock. During a write operation the self - timed RAM will internally generate the proper timing for the write operation. System designers are no longer concerned with meeting write pulse width, data setup and hold relative to the write pulse, address setup and hold relative to the write pulse, and select setup and hold relative to the write pulse. System signal skews in very fast systems routinely cause the write cycle timing to be a critical speed limiting problem. Self - timed RAM operation provides a viable solution.

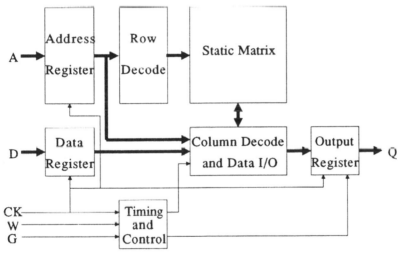

Figure 7.7: Self - Timed SRAM Functional Block Diagram.

The block diagram in Fig. 7.7 illustrates one of several types of self - timed SRAM. Registered inputs and registered outputs are shown, but applications exist for other variations; latched inputs and outputs, registered inputs and latched outputs. There are subtle differences in the timing diagrams for the various types, but the differences are not important to the understanding of self - timed device operation. Self - timed operation is best understood by comparing the timing diagram to the standard SRAM and synchronous SRAM timing. The speed performance of a self - timed SRAM should not be noticeably different than an asynchronous SRAM designed with equal care, using the same process and power budget. The speed enhancement comes in the system, where the self - timed SRAM is more easily used at nearly the data sheet cycle time. For example, a 10 ns ECL I/O asynchronous SRAM may be usable at 15 to 20 ns cycle time in a well designed system (designed with attention to signal skews). A 10 ns self - timed ECL I/O SRAM relieves system level skew problems to the extent that a well designed system could achieve 12 to 15 ns cycle time.

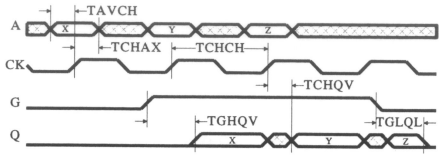

Figure 7.8a: Self - Timed SRAM Read Timing Diagram.

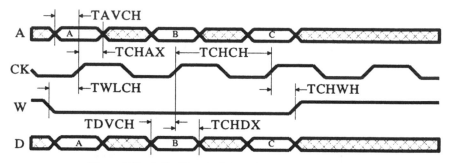

Figure 7.8b: Self - Timed SRAM Write Timing Diagram.

Most notable is the difference in the write cycle timing of the self - timed device. All inputs are registered at the beginning of the cycle, and the on chip timing circuitry completes the write operation within the cycle time. Simplified write pulse width, write setup and hold times (relative to address, data and select) are the result. Data output is held stable for much of the cycle time by the output register, easing system constraints for the data path timing. These differences in timing are the key differences which allow the system designer to design for high speed more easily. Another notable feature is the pipeline - like operation. Devices with output registers are usually designed such that the same clock signal clocks the inputs and the outputs simultaneously. At the time a set of inputs are clocked in, the output from the previous operation is clocked out. This mode of operation causes the RAM to behave as a pipeline with a depth of one stage.

Applications and Benefits. Self - timed SRAM devices are becoming popular for cache memory and also for very high bandwidth applications within CPU's. Applications within the CPU include writable control store, deep register files (for register file windowing) and high speed buffers. In addition to the cache memory itself, they are also useful in cache tag and address translation buffers, associated with cache controller and memory management controller functions. Applications within the CPU are extremely speed sensitive. Control store and register files are in the execution path, and impact the machine cycle time. Most high speed CPU designs are pipelined synchronous architectures, so self - timed RAM fits nicely.

BiCMOS is a logical process technology choice for self - timed RAM primarily because of the focus on very high speed applications. Inside the memory device itself, a self - timed SRAM is similar to any other synchronous SRAM. There is simply more latching, registering and timing done on chip. The same speed and design flexibility issues covered earlier (in section 7.2.1) become even more critical. Timing skews on chip are much more of a concern, and BiCMOS or bipolar ECL circuitry can be more easily designed for minimum skew than CMOS alternatives. In self - timed SRAM's of the 10 ns speed class, skew (on chip) in the range of 100 to 200 ps may become critical. Bipolar ECL self - timed devices will generally be faster but are limited to lower densities by the severe speed - power - density trade-offs inherent in bipolar memory design. BiCMOS provides ECL I/O compatibility at the higher densities some applications require.

7.2.3 Pipelined SRAM

Pipelined SRAM's are self - timed memories which allow cycle time to be shorter than access time. These devices can accept new inputs (i.e. new addresses, data in and control signals) before the access from the prior set of inputs is completed. Pipelined devices allow the shortest cycle time and highest memory bandwidth for a given technology and memory density. They are, however, more difficult to design, and fairly difficult to utilize in applications. The advantage of very short cycle time outweighs the difficulties for some very high speed systems. Most pipelined SRAM devices are custom devices, specified by a given customer for application in a specific machine. This is a relatively new area and over the coming years an increasing variety of devices of this type are anticipated. Fig. 7.9 provides a representative block diagram of one type of pipelined SRAM.

A simplified timing diagram is shown in Fig. 7.10. This diagram does not illustrate the timing parameters in detail; it is intended to show the overall timing concept of cycle time less than access time. Cycle time is set at one half of access time for the illustration. With more complex device architectures it is possible (but increasingly difficult) to develop memories capable of supporting cycle times of one third to one fourth access time with a similar design philosophy. If even shorter cycle time is necessary, it is conceivable to implement a memory device which is internally interleaved using the same concept as board level interleaved architectures (discussed in section 7.2.1).

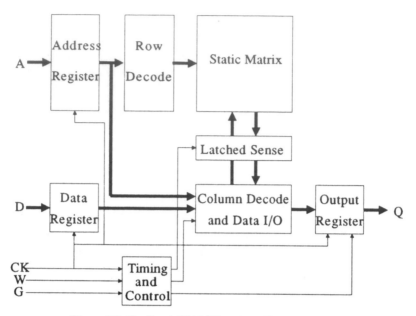

Figure 7.9: Pipelined SRAM Functional Block Diagram.

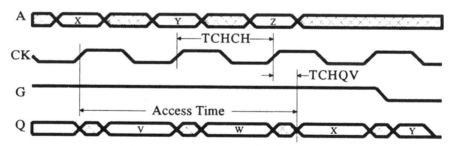

Figure 7.10a: Pipelined SRAM Simplified Read Timing Diagram.

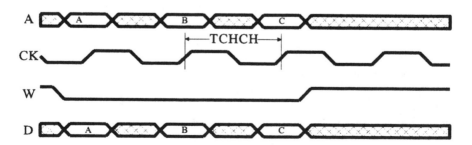

Figure 7.10b: Pipelined SRAM Simplified Write Timing Diagram.

Applications and Benefits. Extremely short cycle time provides very high memory bandwidth to the system designer. Integrating all the registering, timing and controlling logic onto the memory devices increase cost, such that only limited applications are economically feasible. Low density pipelined memory is of interest for applications within the CPU, in the same types of uses as self - timed SRAM. Higher density pipelined SRAM is useful in large second level cache applications. High density pipelined SRAM is also occasionally considered for main memory in large supercomputers.

7.2.4 Latched and Registered PROM

Latched PROM and registered PROM are synchronous memories that stand out because of their applications more than their architecture. Bipolar PROM has been offered with output latches and registers as optional features for many years. Choices of technology (e.g. bipolar, BiCMOS) and of nonvolatile storage elements (e.g. fuse, floating gate MOSFET) are not important in understanding the breadth of applications. It is sufficient to simply be aware that there are a variety of technologies and storage elements which a device designer can choose. Most latched or registered PROM features latches or registers on the data outputs only, as illustrated in Fig. 7.11.

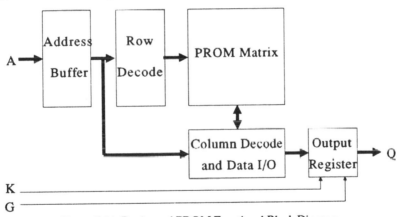

Figure 7.11: Registered PROM Functional Block Diagram.

Figure 7.12 illustrates the relatively straightforward read cycle timing of a typical registered PROM. Only the outputs are registered. Address inputs propagate asynchronously through the decoders and the accessed data is sensed by the sense amplifiers. The clock synchronizes the output transitions to the system timing.

Applications and Benefits. The applications of synchronous PROM are quite varied. Control store (machine level microcode) is one application, but most modern microcoded machines are designed with synchronous RAM to allow writable control store. Common applications for these devices are ASIC logic function blocks. A PROM is quite similar to a PAL (PAL is a registered trademark of AMD/MMI Inc.) in that it is readily programmed to implement an arbitrary logic function. PAL devices are programed to implement SOP (Sum Of Products) boolean equations. PROM can similarly be programmed to implement POS (Product Of Sums) equations. The PROM can be thought of as a bank of fixed AND gates (the decoders) followed by programmable OR gates (the PROM cells and sense amplifiers form a programmable wired OR function). Standard PROM is quite useful for implementing arbitrary asynchronous logic functions. A PROM is sometimes more useful than a PAL because a PROM is fully decoded; providing all possible product terms for a given number of inputs (addresses). PAL type devices do not provide all possible product terms, although they are adequate for many applications. A latched or registered PROM is quite useful in implementing arbitrary synchronous logic functions. Feedback from some of the outputs to some of the inputs implements a simple and versatile programmable state machine.

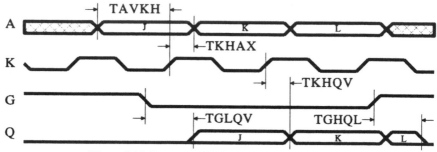

Figure 7.12: Registered PROM Read Timing Diagram.

Few examples of BiCMOS nonvolatile memory devices are available. The relative newness of the process technology is the primary reason for the dearth of programmable BiCMOS memory devices. As the technology matures an increasing variety of nonvolatile memory functions are likely to emerge [7.3]. There are several reasons for this expectation. Higher densities of TTL I/O PROM become more practical in BiCMOS than bipolar processes. ECL system designers have long been hampered by the lack of programmable memory and logic devices. The few available have generally offered very poor speed and functional density compared to the expectations of the ECL system designers. These factors lead to the conclusion that BiCMOS will likely become a common process for high performance nonvolatile memory in the near future.

7.3 Multiport Specialty Memories

Multiport memory devices have two or more data ports including address and control inputs for each data port. Separate I/O SRAM devices are not multiport memories because they do not have two sets of address and control inputs, one for the read outputs and one for the read inputs. Multiport memories may feature asynchronous style interface on each port, synchronous style interface on each port, or even possibly some ports with each style. Common examples of multiport memories include FIFO, dual port SRAM, and multiport register files. FIFO memories are a special case of multiport devices; addresses are generated internally (if the FIFO utilizes a RAM based architecture internally). Multiport devices may incorporate some form of on chip contention arbitration circuitry. Common dual port SRAM memory has on chip arbitration logic which precludes simultaneous accessing from the two ports irrespective of the address. If an operation is in progress on one port when an access is requested on the other port, the second port receives a "busy" response and is forced to wait for the first access to finish. Register file applications generally require the ability to read and / or write at several address locations in the same machine cycle. To support multiple simultaneous accesses, register file memories are architecturally more complex.

7.3.1 FIFO Memories

FIFO memory devices are specialty memories designed to implement a FIFO (First In - First Out) buffer. They are designed to allow simultaneous and asynchronous access from both the read and the write ports. Synchronous style interfaces for each port are most common (e.g. a read clock and a write clock); the asynchronous access refers to the freedom for the timing of one port to be totally independent of the timing for the other port. Most FIFO memories are based on a dual port memory cell (Fig. 7.13). An array of these dual port cells allows multiple simultaneous accesses. While, for example, the read pointer is addressing a given cell to read it, the write pointer can access any other cell to write it. FIFO memories normally have on chip circuitry to preclude writing and reading at the same location at the same time. Simultaneous selection of a cell could occur if the internal logic did not prevent writing when the FIFO is full, and reading when empty.

Typical RAM based FIFO memories contain two separate internal address counters, one for the read address and one for the write address. Most include flag generation logic. Flags are outputs which indicate status. Full, empty and half full flags are found on most new devices; a few devices also offer almost empty and almost full flags in addition. Flags are generated by comparing the read and write address. If the counters are equal the condition is either full or empty, depending upon which counter was incremented to cause equality. Most new FIFO memories today also contain logic to support depth expansion. One organization of FIFO which is common today is 512 x 9. If, for example, a FIFO of 1536 x 18 is desired in an application, both depth and width expansion will be required. Width expansion is straightforward; simply wiring two adjacent devices in parallel supports an 18 bit word. Depth expansion is more complicated because the addressing is done on chip. The second device needs to start writing when the first becomes full, and so on for all devices. Reading also creates the same type of problem. Modern FIFO memories include expansion in / expansion out logic to essentially "pass the baton" to the next device. The system designer must interconnect the devices for proper operation of the depth expansion logic, but the logic is conveniently implemented on chip.

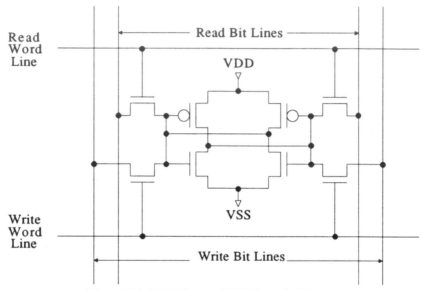

Figure 7.13: FIFO Memory Cell Schematic Diagram.

Expansion mode causes the flags of any one device to have a different meaning than a one deep array. Continuing with the same 1536 x 18 example, the flags from the devices storing bits 0 - 8 will provide the exact same information as the flags from the devices storing bits 9 - 17. Only one set of flags are utilized in width expanded arrays; the other sets are ignored. If only two of the three (deep) full flags are asserted at a given instant, the array is at least two thirds full, and at least one word less than totally full. A few external logic gates are required to derive meaningful full and empty flags for the array. Reset logic is another important logic feature required for proper operation. By resetting each FIFO at power - up, the system ensures that all address counters start at zero (empty condition). Figures 7.14 and 7.15 illustrate functionality and timing of a FIFO.

Applications and Benefits. FIFO memories are most useful in applications where temporary storage of data is required when passing information between two different sections of a system, especially if the data rate capabilities of the two different sections are different. Within a high speed processor, for example, a small FIFO may be useful for buffering the data passed between an ALU (Arithmetic Logic Unit) and an FPU (Floating Point Unit). Deep FIFO's may be desired for buffering information between a very fast CPU and each one of several slower IOP's (Input / Output Processor). The faster CPU need not be stalled waiting for each word of input; instead it receives a large block of input at the CPU's data rate from the FIFO. In essence the FIFO provides a "bucket" for data which may be filled quickly and emptied slowly, or filled slowly and emptied quickly. FIFO memory is usually employed where different data rates complicate communications.

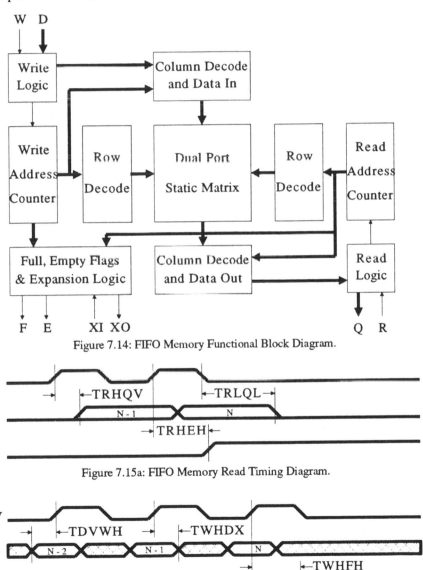

Figure 7.14: FIFO Memory Functional Block Diagram.

Figure 7.15a: FIFO Memory Read Timing Diagram.

Figure 7.15b: FIFO Memory Write Timing Diagram.

Very high speed FIFO memories are desirable for high speed digital communications systems. BiCMOS is an attractive technology for very fast FIFO memory (above 50 MHz) due to the performance achievable in BiCMOS designs. ECL FIFO devices have not been available in the past, forcing ECL system designers to implement less desirable alternatives with SRAM and logic. Dual port memory cells implemented in traditional bipolar ECL become prohibitively large and expensive. BiCMOS offers the promise of ECL I/O FIFO memory because the dual port cell is efficiently implemented in a CMOS core, and the ECL I/O circuitry is easily supported.

7.3.2 Dual Port SRAM

Common dual port SRAM devices are constructed using standard SRAM cells. They can not support multiple simultaneous accessing (as the FIFO can). Dual port SRAM is normally designed with on chip arbitration logic, which provides a "busy" response to the access request if another access is in progress on the other port. Many dual port memories in use today feature asynchronous RAM - like interfaces for each port (with an added busy signal), but a few feature synchronous interfaces with address latches and clocks [7.4].

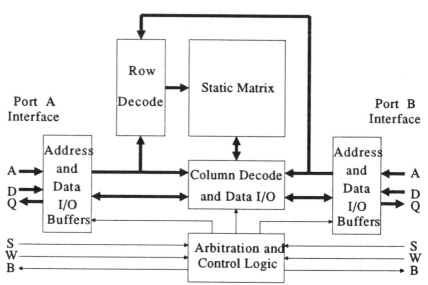

Figure 7.16: Dual Port SRAM Functional Block Diagram.

Dual port devices require some special rules for constructing larger arrays, just as FIFO memories do. Dual port devices can be easily configured for deep arrays, but the difficulty comes in width expansion for wide words. An example array of 4K x 32 is desired for shared data in a multiprocessor system. Sixteen 1K x 8 devices will be needed. A difficulty arises because there is arbitration circuitry on each device. When any given 32 bit word is accessed the designer must be sure that all 4 devices accessed will allow port A access, for example, and none must allow port B access. If both processors make a request at nearly the same instant, it could result in different decisions by the different arbiters on each device. Such a condition is clearly intolerable. Most devices overcome this by providing some type of a slave feature. This allows the system designer to designate only one device (in each bank of depth) as a master, with all other devices (in that same bank of depth) slaved to follow the master's arbitration decision.

Port A

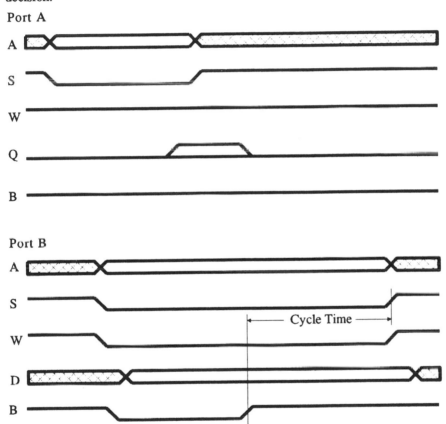

Figure 7.17: Dual Port SRAM Simplified Timing Diagram.

Some dual port SRAM provide flags or semaphores which are helpful in signalling when a message or block of data is ready to be read from the other port. Flags assist the system designer in implementing interrupt driven message passing schemes, as opposed to polled schemes. Arbitration between ports, or contention resolution as it is also called, is illustrated by the timing diagram. If one port is busy the other must wait for the first to complete the access. Except for the busy signal, the dual port SRAM acts just like a standard SRAM from the viewpoint of either port. Figure 7.17 illustrates port A responding to a read access request, causing the busy response to the port B write access request. The delayed cycle completes a cycle time after the busy condition clears.

Applications and Benefits. Dual port memory applications are primarily found in multiprocessor systems, where the dual port devices are used for shared memory allowing message or data passing between processors. The advantage of dual port memory over FIFO memory for this type of application is the ability to randomly access the information in the dual port, where a FIFO is restricted to sequential accessing. FIFO memory, alternatively, offers the benefit of simultaneous and asynchronous access from either port, which most dual port memories can not support. A few dual port devices have been developed which utilize the same type of memory cell as the FIFO, and therefore can support simultaneous and asynchronous access from either port.

7.3.3 Multiport Register Files

Devices with several ports are very useful as register file memories. Register files need to support multiple and simultaneous accesses so a simple SRAM cell is unacceptable. A five port device, for example, is normally designed with five select lines (rows in the matrix) able to select any given cell. Figure 7.18 illustrates such a cell. Five port cells are no different in concept than the two port cell used in a FIFO, but are much larger in practice due to the additional word lines, select gates and bit lines.

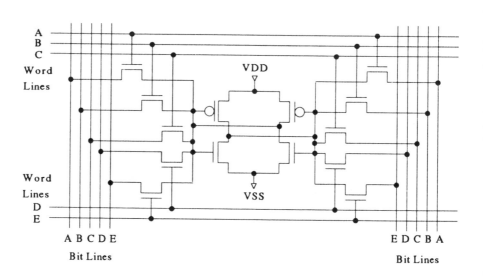

Figure 7.18: Five Port SRAM Cell Schematic Diagram.

Register file devices are often designed with specific ports dedicated for reading and other ports dedicated for writing. Figure 7.19 shows an example five port device which features two write ports and three read ports.

Figure 7.20 is a simplified timing diagram for a five port SRAM with asynchronous style interface. It illustrates the capability to support several simultaneous accesses. Detailed timing requirements are not illustrated, since they are the same as a standard asynchronous SRAM.

Two Write Ports   Three Read Ports

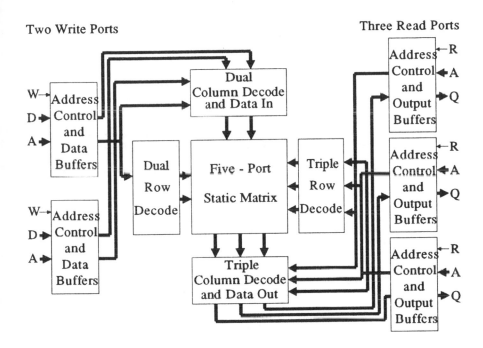

Figure 7.19: Five Port SRAM Functional Block Diagram.

Applications and Benefits. High speed CPU architectures usually include several register file memory devices. The register file is a relatively low density storage for temporary local buffering of operands and results. Register files are often multiport devices because the CPU usually requires two operands and generates one result in each machine cycle. Two read ports and one write port are often dedicated to the CPU. It is an advantage to provide an additional read and write port for passing data between the register file and the cache or main memory. A memory load or store may be accomplished in the same cycle as an arithmetic or logic operation with this type of device.

Register file memories are required to be extremely fast because they are usually in the CPU critical speed path. ECL register files have historically been implemented in costly and high power bipolar processes to achieve the highest possible speed. BiCMOS offers a more cost effective and lower power option due to the reduced cell size and reduced cell power consumption resulting from a CMOS array.

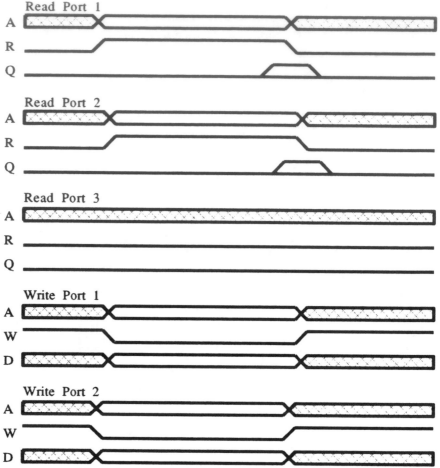

Figure 7.20: Five Port SRAM Simplified Timing Diagram.

7.3.4 Video DRAM

Video DRAM is another type of multiport memory in common use today. Video DRAM can be read and written as a normal DRAM, and also contains a serial shift register used for reading out a high speed serial bit stream [7.5, 7.6]. Recent generation video DRAM also include special operating modes specifically intended to simplify pixel manipulations in video systems. A video DRAM functional block diagram is shown in Fig. 7.21.

Video DRAM supports read and shift serial port cycles while random access cycles occur on the RAM port, as illustrated in the simplified timing diagram shown in Fig. 7.22. (The timing diagram illustrates MOS / TTL I/O levels; all standard video RAMs are MOS / TTL I/O compatible at this time.) The video DRAM accomplishes seemingly simultaneous access on each port by filling a large serial shift register from the memory array. One complete row is usually transferred from memory into the shift register. While the serial shift register provides video data at the serial port, the RAM port allows accesses to the RAM array for modifying the video information. The cell matrix does not require dual port cells since there is no need for true simultaneous accesses to the matrix.

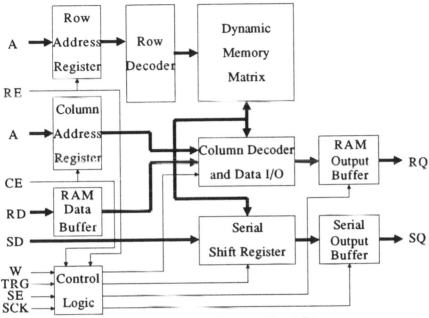

Figure 7.21: Video DRAM Functional Block Diagram.

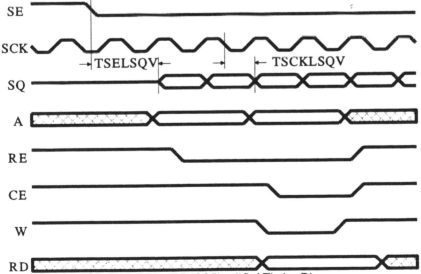

Figure 7.22: Video DRAM Simplified Timing Diagram.

Applications and Benefits. Video DRAM memory is highly specialized for video frame buffer applications. The serial shift register feature is quite useful as it provides the serial bit stream at the video clock rate for the video DAC (Digital to Analog Converter). Recent generation video DRAM also include several operating modes which are convenient for implementing video functions such as reverse field, scrolling, etc. Some of the features being offered on recent generation devices are listed below:

Examples of Video DRAM Specialty Features

> Masked Write Transfer Cycle
> Split Write Transfer Cycle
> Read Transfer Cycle
> Load Write Mask Register
> Masked Data Write Cycle
> Flash Write Cycle With Mask
> Block Write Cycle With Mask
> Load Color Register

The variety and number of video specific features in the list is strong evidence of the application specific nature of video DRAM. Video applications continue to progress in resolution and in the number of colors able to be displayed at any given moment. The expanding market demands increasing performance and features from future video DRAM. Over the coming years BiCMOS may become a popular choice for Video DRAM, to support the demands of increased speed (video shift rate) and possibly a migration to ECL I/O for the serial port as speeds become impractical for TTL I/O levels.

7.4 Cache Application Specific Memories

There are a few specialty memory types which are often found in cache memory applications; cache tag memory and CAM (Content Addressable Memory). Cache tag memories are SRAM with on chip data comparators. Some also include a flash clear function. CAM is a type of memory which responds to a data word by indicating whether the word matches any stored word or not. CAM is often found in cache and MMU applications, and is attractive for a few other applications. Extremely high cost per bit prevents it from being economically feasible for most applications.

7.4.1 Cache Tag Memories

Cache tag memories are SRAM devices with comparators on chip. The tag RAM supports normal read and write cycles, as well as a read and compare operation. During a read operation the data output from the addressed location is compared with a word of data supplied externally. If there is a match (each bit matches) a match output signal is asserted. During a read cycle the data input pins are otherwise unused, so many tag RAM designs use the data input pins as the inputs for the data to be compared. Some tag RAM devices also include a flash clear feature [7.7]. The purpose of the flash clear is to reset the contents of the entire tag RAM so that no location in memory can generate an unintentional match when compared with the match data inputs. Figure 7.23 illustrates the functional block diagram of a typical cache tag SRAM, and Fig. 7.24 provides a simplified timing diagram illustrating the read and compare operation.

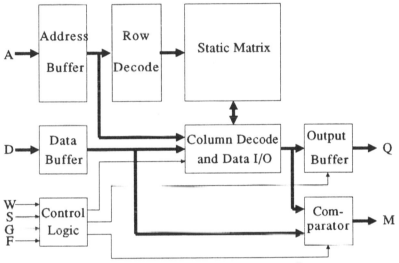

Figure 7.23: Cache Tag SRAM Functional Block Diagram.

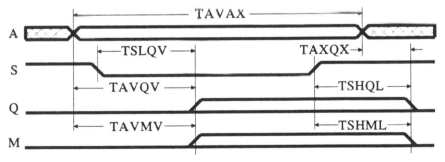

Figure 7.24: Cache Tag SRAM Simplified Timing Diagram.

Applications and Benefits. Cache tag SRAM enhances cache memory system speed since on chip comparators are faster than external comparators. Circuit board wiring delays and the additional capacitance and inductance inherent with discrete comparators give a noticeable speed benefit to the on chip solution of the tag RAM. A standard SRAM followed by an external comparator results in additional delay of about 4 ns in an ECL system, and more than 10 ns in a TTL system. Most architectures will result in an access of the tag RAM for every memory reference. The match indicates that the referenced data is currently available in the cache. If there is not a match, the data must be fetched from main memory. The speed of the tag RAM is critical because it is one of the pacing items in many operations. Flash clear functionality enhances the speed of the system in handling a context switch. When the context is switched all contents of the cache need to be invalidated, and the cache will be refilled by subsequent misses generated by memory references from the new context. It would be quite time consuming if the machine must individually write to every tag location to clear it. Some tag RAM memories are designed to support a flash clear of all memory locations in only one or a few cycle time periods.

7.4.2 Content Addressable Memory

CAM is a highly specialized and unusual form of memory which is extremely useful in high speed lookup table applications. CAM can accept a data word as an input and, if there is a matching data word stored in the array, the CAM will return a match signal along with the address of the matching location [7.8]. CAM can also be read and written just as normal SRAM. CAM requires a considerably more complex cell than standard SRAM because it implements a hardware bit by bit comparison within every cell, as seen in the example cell shown in Fig. 7.25.

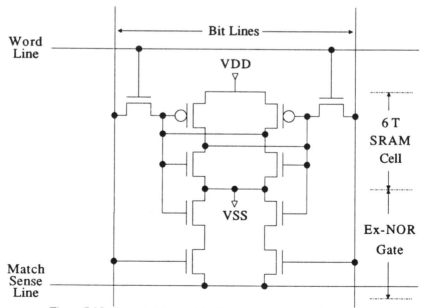

Figure 7.25: Content Addressable Memory Cell Schematic Diagram.

Figure 7.27 is a simplified timing diagram for a content addressable memory. It illustrates the key function of a CAM; the ability to input a data word and result in a match output if a match is found. Some CAM devices also include the ability to mask certain portions of the data word, allowing the comparison for a match to be made for only a subset of the bits stored in each word.

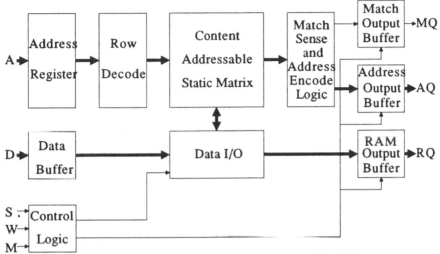

Figure 7.26: Content Addressable Memory Functional Block Diagram.

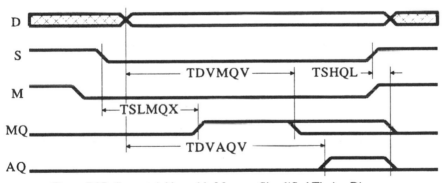

Figure 7.27: Content Addressable Memory Simplified Timing Diagram.

Applications and Benefits. CAM is very unique in that it quickly checks all locations for matching data. Tag RAM, in contrast, can check for a match only at the location addressed. CAM is very useful in cache and MMU applications, for example, as a lookup table supporting virtual to real address translations (a page map table). High cell complexity and the extra match lines in the array cause much larger cell area than standard SRAM, resulting in relatively low densities and very high cost per bit. BiCMOS is a desirable process for CAM devices even more than for standard SRAM. The large CAM cell causes relatively long and heavily loaded lines in the array, as compared to standard SRAM. The match data input lines are in the speed critical path, and may be quickly driven by bipolar drivers. A cell implementation with CMOS storage (six transistor, or four transistor / two resistor) and a bipolar match output driver might provide superior performance with acceptable power and density. The benefits are similar to all other memories; as performance goals are increased, the design flexibility of BiCMOS increases in value.

7.5 Imbedded Specialty Memory

Possibly the ultimate in specialty memory is imbedded memory; memory imbedded within an ASIC device (e.g. gate array, standard cell) [7.9]. By imbedding application specific memory within an ASIC or full custom logic device several advantages may be realized. The memory can be organized and configured in a way that best suits the application. Writable control store imbedded within a CPU logic block, for example, can be as wide as the machine requires without penalty in pincount or the time lost accessing off chip control store. Imbedded register files are also attractive choices due to the relief from pincount issues (multiport devices have high pincount) and wiring delay. Both applications are often small enough in total bit count that imbedded memory is worth consideration. Cache memory controller and MMU applications also require relatively small amounts of specialized memory (e.g. CAM) so the same benefits may be realized with imbedded memory implementations [7.10].

Benefits of imbedded memory are not achieved without a price. There are several costs to be considered, and weighed against the potential performance gains. Not all the costs involved are easily quantified. Time to market is an example of a cost to be considered. The current state of tools available for such development effort is far better than only a few years ago, but still falls short of customer desires. Many ASIC vendors provide macros for easy design of memory imbedded in standard cells and gate arrays. Such tools give a starting point, but not usually the architecture best suited for the specific application. Additional logic may be required to create registers, timing generators, and similar functions needed for each specific application.

Testing imbedded memory is also very difficult. Discrete memory devices are relatively easily tested for functionality, performance, pattern sensitivity, and more. Imbedded memory is more difficult to test at high speed to ensure adequate margins. It is also difficult to test with complex memory test patterns. High density ASIC devices are often designed with scan diagnostic paths to assist in functional testing, but serial scan is inadequate when attempting to pattern test an imbedded memory at full rated speed.

Other more subtle and insidious difficulties can arise with imbedded memory. Reliability is a major concern of most memory vendors. Reliable devices start from reliable designs and processes. Many ASIC vendors lack experience with, for example, soft error rate minimization techniques employed in memory devices. BiCMOS can help here too. Chapter 4 (section 4.3) discussed the benefits of BiCMOS in control of alpha particle induced soft errors in SRAM's. BiCMOS memory has demonstrated alpha particle tolerance superior to current generation bipolar and CMOS SRAM cell implementations.

When considering ASIC devices with imbedded memory, a complete system perspective is necessary to properly weigh the benefits and costs in light of the overall system performance goals (not just speed performance) and cost target.

7.6 Economics Of Specialty Memories

It is fundamental to understand that the focus of specialty memory is to bring value, usually in the form of higher speed, to the system application. It really does not matter if a specialty memory is the same speed as a standard RAM on the device specification sheets. Performance actually achieved in the application is the acid test. It is conceivable that the additional logic (e.g. registers) within a specialty memory could actually cause the device to be slightly slower than a competitive generic device. For certain applications the device may yet offer the benefit of improved system speed.

Increased costs of increased specialization need to be considered. Standard DRAM and SRAM are widely sourced, very competitive, and very inexpensive per bit. Specialty memories cover a wide range in pricing per bit. Relatively simple devices such as latched and registered SRAM costs less than more esoteric specialty memory types, but may not provide the system performance desired. Video DRAM is manufactured in high quantity, and is competitive enough to be reasonably priced in spite of the added complexity. Devices with very large memory cell structures and low volume demands in the marketplace, such as CAM and multiport register file memory, are at the high end of the cost per bit spectrum. Learning curve cost reductions help drive costs down more rapidly for the more popular specialty devices, but prices decline more slowly for the more unusual devices.

System architects and design engineers must analyze the choices between standard memories, specialty memories and imbedded memories. They should consider not only component costs, but a long list of less tangible costs such as development costs, testing costs, repair costs, and more. They need to consider the goals of their design, and the importance of system performance to the success of the system in the market. Each system has unique requirements resulting in decisions unique to each system.

7.7 Conclusions

Specialty memories cover a broad range of device types intended to provide benefit for a wide variety of specific applications. There is a full spectrum available, from standard memory to specialty memory to custom memory and imbedded memory. Increasing specialization brings increasing performance at increasing cost. As technology continues to make strides in density and performance, the trade-offs between standard, specialty, custom and imbedded memory constantly change. The combination of high packing density and high performance achievable with high speed BiCMOS processes makes specialty memory an attractive area of focus for BiCMOS memory designers.

References - Specialty Memories

7.1] Dataquest Inc., MOS SRAM Forecast, pp.15, July 1988

7.2] R. Kertis, D. Smith, T. Bowman "A 12ns 256K BiCMOS SRAM" ISSCC Digest of Technical Papers, pp.186-187 and 362-363, February 1988

7.3] C. Sung, P. Sasaki, R. Leung, Y. Chu, K. Le, G. Conner, R. Lane, J. deJong, R. Cline "A 76MHz Programmable Logic Sequencer" ISSCC Digest of Technical Papers, pp.118-119 and 306, February 1989

7.4] F. Barber, D. Eisenberg, G. Ingram, M. Strauss, T. Wik "A 2Kx9 Dual Port Memory" ISSCC Digest of Technical Papers, pp.44-45 and 302, February 1985

7.5] R. Pinkham, D. Russel, A. Balisteri, T. Nguyen, T. Herndon, D. Anderson, A. Mahta, H. Sakurai, S. Hatakoshi, A. Guillemaud "A 128Kx8 70MHz Video RAM with Auto Register Reload" ISSCC Digest of Technical Papers, pp.236-237 and 384, February 1988

7.6] K. Ohta, H. Kawai, M. Fujii, T. Nishimoto, S. Ueda, Y. Furuta "A 1Mbit DRAM with 33MHz Serial I/O Ports" IEEE Journal of Solid State Circuits, pp.649-654, October 1986

7.7] A. Suzuki, S. Yamaguchi, H. Ito, N. Suzuki, T. Yabu "A 19ns Memory" ISSCC Digest of Technical Papers, pp.134-135 and 367, February 1987

7.8] H. Kadota, J. Miyake, Y. Nishimichi, H. Kudoh, K. Kagawa "An 8Kbit Content Addressable and Reentrant Memory" IEEE Journal of Solid State Circuits, pp.951-957, October 1985

7.9] Y. Sugo, M. Tanaka, Y. Mafune, T. Takeshima, S. Aihara, K. Tanaka "An ECL 2.8ns 16K RAM with 1.2K Logic Gate Array" ISSCC Digest of Technical Papers, pp.256-257, February 1986

7.10] J. Cho, J. Kaku "A 40K Cache Memory and Memory Management Unit" ISSCC Digest of Technical Papers, pp.50-51 and 302, February 1986

Appendix A

JEDEC / IEEE Standard Timing Diagram Nomenclature

Pin names are abbreviated for convenience in diagrams, especially in timing waveform diagrams. Many of the more common abbreviations are listed below:

Abbreviation	Signal / Pin Name	Comments
A	Address	individually numbered 0-N
D	Data input	individually numbered 0-N
Q	data output	individually numbered 0-N
E	chip Enable	select and power control
S	chip Select	no power control
W	Write enable	
G	output enable	output control only
C	input Clock	inputs only
K	output clock	outputs only
CK	in / out clock	clocks inputs and outputs
RE	Row Enable	row address latch enable
CE	Column Enable	column address latch enable

Only a few more abbreviations are needed (along with the above pin / function abbreviations) to have a very concise description for each timing parameter. The other abbreviations necessary are:

Abbreviation	Definition	Comments
T	Time	
H	High	transition to high
L	Low	transition to low
V	Valid	transition to valid
Z	high Z	transition to 3 state
X	undefined	transition to changing

Unambiguous timing signal names become very straightforward with these few abbreviations concatenated in the following way:

T S L Q V   =   Time from Enable going Low to Output going Valid
              =   Chip Enable Access Time

Some of the more common timing parameters are listed below:

TSLQV	select access
TAVQV	address access
TAVAX	cycle time (asynchronous)
TCLCL	cycle time (synchronous)
TWLWH	write pulse width
TAVWL	address setup time to write
TWHDX	data hold time from write
TSHQL	output disable time (ECL I/O)

Conventions used in the timing waveforms are shown below:

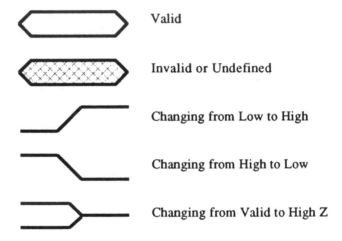

Valid

Invalid or Undefined

Changing from Low to High

Changing from High to Low

Changing from Valid to High Z

Chapter 8

Analog Design

H.-S. Lee, Massachusetts Institute of Technology

8.0 Introduction

In the early days of integrated circuits, analog circuits were built primarily in bipolar technologies. Much of the design methodology for analog integrated circuits evolved from the bipolar technologies with vertical *npn* transistors and lateral *pnp* transistors. While bipolar analog integrated circuits continued to develop, NMOS technologies were becoming popular in digital integrated circuits for their high packing density. For analog applications, despite the inherent drawbacks of NMOS circuits such as low gain, difficult level shifting, and large offset voltages, many clever circuit design techniques were invented to overcome the shortcomings, and take full advantage of NMOS technologies [8.1-8.4].

The most important distinction between bipolar and MOS integrated circuits is the type of signals. In bipolar analog circuits, signals are often carried in the form of currents because bipolar transistors have good linearity in the current domain. In NMOS technologies, infinite input resistance of MOS transistors combined with high quality MOS capacitors make it ideal for charge to be the signal. Many of the important circuit design innovations in NMOS technologies such as switched-capacitor filters and charge-redistribution A/D converters employ charge as a signal. [8.1-8.2]

CMOS technologies followed NMOS technologies in digital circuits for the obvious advantage of low power dissipation. For analog circuits, CMOS technologies removed some of the limitations of NMOS technologies. Due to increased gain and simple level shifting provided by CMOS technologies, both accuracy and frequency response of analog circuits were improved. It became possible to build higher performance analog circuits that were compatible with dense digital logic circuits [8.5-8.9].

Although CMOS technologies greatly improved the MOS analog circuit performance, for high-performance analog circuits, bipolar technologies still have many advantages. Bipolar transistors have much higher transconductance than MOS transistors at a given bias current level, which translates into higher gain, speed, and lower noise. However, incompatibility with charge circuit techniques such as double correlated sampling and switched capacitor techniques has limited the application of bipolar technologies. This requires that for high-accuracy applications thin film resistors and laser trimming be used for the correction of offset and other errors.

BiCMOS technologies, which are emerging as the next generation technology for digital circuits, can greatly benefit analog circuits by making it possible to combine high performance bipolar transistors with MOS circuit design techniques. There have been a few BiCMOS technologies in the past, utilizing triple diffused structure. In such technologies, the performance of bipolar transistors is severely limited due to the high collector series resistance, high parasitic capacitance, and low cut-off frequencies. This chapter assumes modern high performance BiCMOS technologies with optimized *npn* bipolar transistors[8.10-8.12]. (also see Table 8.1 in the next section.)

Although BiCMOS technologies open up endless possibilities for new analog design techniques, not much information is yet available in the literature. Some of the possibilities for new BiCMOS analog circuit design techniques are examined in this chapter. Section 8.1 summarizes device requirements for analog or analog/digital BiCMOS processes. In Section 8.2, electrical properties of MOS and bipolar transistors are compared for analog applications. In Section 8.3, analog subcircuits such as current sources and basic amplifiers are described. In Section 8.4, operational amplifier design techniques are examined, and in Section 8.5, BiCMOS analog subsystems are discussed.

8.1 Devices in Analog/Digital BiCMOS Process

High performance analog circuits impose a different set of requirements on the device characteristics from that of digital circuits. For analog-only applications, the process should be optimized for that purpose. However, since many systems have not only analog circuits but also digital circuits on the same chip, a compromise should be made for both kinds of circuits. The circuit design techniques discussed in this chapter assume such a compromised analog/digital BiCMOS process, although many of the design techniques are applicable to digital BiCMOS processes.

The most important consideration on the process is the power supply voltages. High-speed, high-density digital circuits require short channel, thin oxide MOS transistors, limiting the power supply voltage to 5 V or less. On the other hand, for high dynamic range and design flexibility, analog circuits require high supply voltages. Generally, 10 V supply voltage is preferred for analog circuits, while the popular 5 V supply voltage would be adequate for low to medium accuracy (corresponding to 8-12 bit accuracy) analog circuits. The following device requirements are based on a 10 V operation.

Parameter	Symbol	Typical Value	Unit
Forward current gain	β_F	80	
Early Voltage	V_A	50	V
Collector resistance	r_c	50	Ω
Emitter resistance	r_{ex}	10	Ω
Base resistance	r_b	200	Ω
E-B junction capacitance	C_{je0}	70	fF
C-B junction capacitance	C_{jc0}	50	fF
C-S junction capacitance	C_{js0}	300	fF
C-B junction breakdown voltage	BV_{CBO}	25	V
C-E breakdown voltage	BV_{CEO}	11	V
E-B junction breakdown voltage	BV_{EBO}	6	V
Unity current gain frequency	f_T	5	GHz

Table 8.1 Typical *npn* transistor parameters in 10 V analog/digital process. The emitter size is 4 μm by 4 μm.

8.1.1 Bipolar Transistors

For a 10 V operation, BV_{CEO} over 10 V is generally preferred, although not absolutely mandatory. In actual circuits, BV_{CEO} becomes important only when a common-emitter amplifier is driven from a high-impedance source, and V_{CE} is large. This situation can be avoided by adding an emitter follower at the input, or cascoding the collector of the common-emitter amplifier.

An Early Voltage of 25 V or higher is desired for high gain. This gives an intrinsic gain (the voltage gain of a common-emitter amplifier with an ideal current source load) of 1,000 or higher at room temperature. A typical value of the Early Voltage for a 10 V analog/digital BiCMOS process is 50 V. f_T of at least 5 GHz is preferred for high-frequency circuits. Also, all the parasitic capacitances should be minimized, for they cause parasitic poles that can limit the frequency response of the circuit.

Collector series resistance must be minimized to reduce Miller effect, and to prevent the transistor from being saturated. Collector resistance less than 100 Ω is satisfactory for most applications. Base resistance adds thermal noise, and also limits the frequency response of a common-emitter amplifier, and thus should be minimized. High frequency bipolar transistors require thin base resulting in large base resistance. A trade-off should be made between the frequency response and base resistance. Base resistance of 200 Ω and f_T of 5 GHz would be a good compromise. The emitter is usually heavily doped so that emitter resistance generally is not troublesome. β_F of 50-200 is often quite satisfactory. Table 8.1 summarizes the *npn* transistor parameters in typical analog/digital BiCMOS processes.

Parameter	Symbol	Typical Value NMOS PMOS		Unit
Substrate doping	N_A	10^{15}	10^{16}	cm^{-3}
Gate oxide thickness	t_{ox}	30	30	nm
Channel mobility	μ_n, μ_p	600	250	$cm^2/V-sec$
Minimum drawn gate length	L	2	2	μm
S/D junction depth	X_j	0.2	0.4	μm
S/D lateral diffusion	L_D	0.15	0.3	μm
S/D overlap capacitance	C_{OL}	0.3	0.6	$fF/\mu m$
Threshold voltage	V_T	0.7	-0.7	V
S/D junction capacitance	C_{j0}	0.1	0.2	$fF/\mu m^2$
S/D junction sidewall capacitance	C_{jsw0}	0.5	1.5	$fF/\mu m$

Table 8.2   Typical MOSFET parameters in 10 V analog/digital BICMOS process.

8.1.2 MOS Transistors

To withstand 10 V reliably, the gate oxide thickness would have to be at least 30 nm, which would compromise the speed of MOS transistors. However, 10 V capability provides more dynamic range and design flexibility for analog applications. Therefore, a trade-off must be made between high voltage capability and speed depending on the applications. Likewise, drain-to-source and field breakdown voltages of more than 10 V are required. Minimum gate length of 2 μ would be acceptable, except in situations where the drain-to-source voltage is large, or where high output resistance is desired, longer gate lengths can be used. As is illustrated in Section 8.2.1, the output resistance is roughly proportional to the gate length.

A low back-gate effect for the transistors which have a common substrate (NMOS transistors in an n-well process, and PMOS transistors in a p-well process) is desirable, because the back-gate effect often reduces power supply rejection, and also limits small-signal gain of a source-follower amplifier (to about 0.8 to 0.9).

A very important but often overlooked problem in MOS transistors is threshold voltage hysteresis due to oxide traps [8.13]. When MOS transistors are subject to a gate-to-source voltage excursion of a few volts, even for a very short period of time (a few microseconds), the traps in the gate oxide within the tunneling distance from the silicon surface are charged quickly, shifting the threshold voltage momentarily. When the gate voltage returns to the normal condition, the traps emit electrons back into the silicon, but with a much slower rate. The result is a hysteresis-like behavior in threshold voltage, which ranges from under 100 μV to a few millivolts, depending on the process and stress conditions. This can be a severe limitation in high-accuracy analog circuits, and must be minimized. For example, if an A/D converter with 10 V full-scale voltage is implemented in a process with 1 mV hysteresis, the accuracy would be limited to 11 bits. Therefore, for high accuracy circuits, the process has to be carefully designed to reduce the trap densities near the interface.

The summary of typical 10 V MOS transistor parameters in analog/digital BiCMOS processes is shown in Table 8.2.

8.1.3 Capacitors

High-quality capacitors are the critical part of many analog integrated circuits. "Free" capacitors such as metal-to-poly capacitors with deposited low temperature oxide should be limited to low accuracy applications (below 10 bits) due to their poor matching properties and large dielectric absorption.

Poly-to-poly capacitors or poly-to-n+ capacitors are usually high quality capacitors with good ratio matching (typically 0.1 % to 0.01 % ratio matching), low voltage and temperature coefficients (lower than 50 ppm/V and 30 ppm/ °C respectively), and minimal dielectric absorption. In some applications, the bottom plate junction leakage current in poly-to-n+ capacitors can limit the performance, in which case poly-to-poly capacitors are preferred.

In most analog applications, rather thick (50 nm or thicker) oxide is desired for capacitors except when a large capacitance value is needed such as in on-chip bypass capacitors and frequency compensation capacitors. Thicker oxide provides better matching per capacitance, and lower voltage and temperature coefficient [8.14].

8.1.4 Other Devices

Lateral *pnp* transistors are usually "free" devices, and as such, they have poor performance. When the lateral *pnp* transistor is forward-biased, the vertical parasitic substrate *pnp* transistor is forward-biased also. The emitter current is split between the lateral and the vertical transistors, giving low α_F (on the order of 0.5). However, β_F can be as high as 100, and the Early Voltage is usually high. f_T of lateral *pnp* transistors is on the order of 10 MHz, and must be avoided in the high-frequency signal paths. The lateral *pnp* transistors, however, can be quite useful as current sources.

Due to many diffusion layers associated with BiCMOS processes, different types of resistors are available. *N* or *p*-well resistors have high sheet resistance on the order of 1kΩ/square, and can be useful in implementing large-valued resistors. However, due to low doping, these resistors have very large temperature coefficients (2500-3000 ppm/ °C), and exhibit significant voltage dependence caused by the depletion of the well under reverse bias. Non-silicide ploy, n or p source/difusion resistors have better temperature coefficient (500-1500 ppm/ °C). Voltage dependence in usually negligible in these resistors, and they also match quite well (8-9 bits). Therefore, it is advisable to use these resistors for precision applications. Generally, however, ratio matching of resistors is not as good as that of capacitors. The shortcoming of these resistors is the low sheet resistance (20-100 Ω/square). Base pinch resistors can also be used as high sheet resistance material, although it is difficult to control the sheet resistance.

8.2 Comparison between Bipolar and MOS Transistors

Bipolar and MOS transistors have many different aspects that must be understood before the advantages of both devices can be exploited while avoiding the shortcomings of either device. In the following sections, comparisons between bipolar and MOS transistors are made in terms of performance parameters relevant to analog circuit applications.

8.2.1 Intrinsic Gain

The "intrinsic" or "open circuit" gain is defined as the small-signal low frequency voltage gain in the common-emitter or common-source configuration with an ideal current source load. The intrinsic gain is the maximum voltage gain that can be obtained from the device, and therefore is a significant parameter in analog circuits. In particular, in operational amplifiers where no more than two gain stages are allowed before frequency compensation becomes difficult, as much gain per gain stage is desired. In the closed loop configurations of operational amplifiers, the error due to finite gain is inversely proportional to the open loop gain. Thus, the intrinsic gain can be considered to be the parameter which sets the upper boundary of the accuracy in the design space.

Consider the small signal model for an npn transistor shown in Fig. 8.1. Among the parameters shown in Fig. 8.1, r_μ is negligible in most transistors, and r_{ex} is typically on the order of a few ohms, and can also be neglected. The intrinsic gain is calculated to be

$$a_o = \frac{r_\pi}{r_b + r_\pi} g_m r_o \qquad (8.1)$$

where

$$g_m = \frac{q I_C}{kT} \text{ , and } r_o = \frac{V_A}{I_C} \qquad (8.2)$$

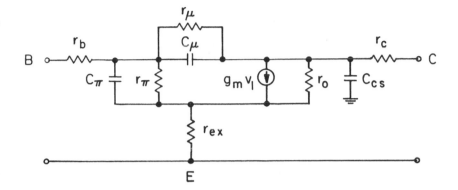

Fig. 8.1 Small-signal equivalent circuit for a bipolar transistor.

Typically, $r_\pi \gg r_b$, and Eqn. (8.1) can be reduced to

$$a_o \approx \frac{qV_A}{kT} \tag{8.3}$$

From Eqn. (8.3), it is readily seen that the intrinsic gain of a bipolar transistor depends only on the Early Voltage and temperature. It is important to note that the intrinsic gain is independent of the size or shape of the transistor, and of the collector current. For a typical Early Voltage of 50 V, the intrinsic gain is approximately 2,000 at room temperature.

In an NMOS transistor in strong inversion biased in the saturation region, the intrinsic gain can be calculated in the similar manner from the small-signal equivalent circuit in Fig. 8.2.

$$a_o = g_m r_o \tag{8.4}$$

where

$$g_m = \left[\frac{2\mu_n C_{OX} W I_D}{L} \right]^{\frac{1}{2}} \tag{8.5}$$

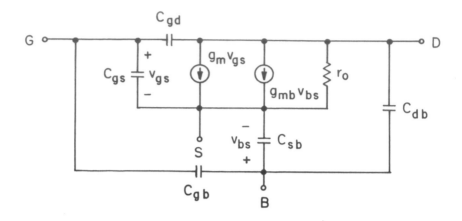

Fig. 8.2 Small-signal equivalent circuit for an NMOS transistor.

and

$$r_o = \frac{1}{\lambda I_D}$$

(8.6)

λ is a parameter which is equivalent to the inverse of the Early Voltage in bipolar transistors. λ depends on process parameters such as the oxide thickness and substrate doping, and is approximately inversely proportional to the gate length [8.16]. A typical value of λ is 0.05 V^{-1} for $L = 2\mu m$. From Eqns. (8.4) through (8.6), the intrinsic gain can be shown to be

$$a_o = \frac{1}{\lambda} \left[\frac{2\mu_n C_{OX} W}{L I_D} \right]^{1/2}$$

(8.7)

As is shown in Eqn. (8.7) in contrast to bipolar transistors, the intrinsic gain of an MOS transistor depends on process parameters, device sizes, and the drain current. One thing that should be noted in Eqn. (8.7) is that the intrinsic gain increases with increasing gate length, because λ is approximately inversely proportional to L, while transconductance is proportional to $1/\sqrt{L}$. Also, the gain increases with decreasing bias current. At very low currents, the transistor is in weak inversion, and the carrier transport is mostly due to diffusion. Therefore, in weak inversion, an MOS transistor behaves much like a bipolar transistor, exhibiting an exponential dependence of drain current on the gate-to-source voltage. In this case, it can be shown that the intrinsic gain is constant, as in bipolar transistors. Although relatively high gain (up to 1,000) can be achieved at low currents, the frequency response is poor at low currents as discussed in the next section.

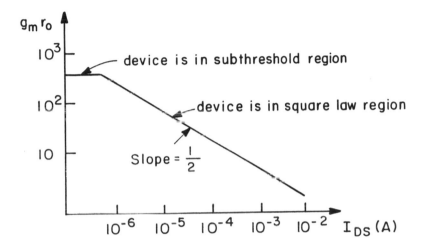

Fig. 8.3 Plot of intrinsic gain vs. the drain current in an NMOS transistor.

A plot of typical intrinsic gain of an MOS transistor as a function of drain current is shown in Fig. 8.3. In strong inversion, intrinsic gain of 50 to 100 is common in typical size transistors, which is significantly lower than that of bipolar transistors.

8.2.2 Frequency Response

Frequency responses of bipolar and MOS transistors can be compared by the cut off frequency ω_T. $\omega_T = 2\pi f_T$ is defined as the angular frequency at which the current gain of the device drops to unity.

In bipolar transistors, ω_T is approximately constant at medium current where the capacitance is dominated by the base diffusion capacitance. ω_T falls off at high current due to high-level injection effects, while it decreases at low current because of parasitic capacitances. At medium current levels, the cut off frequency is approximately the inverse of base transit time.

$$\omega_T \approx \frac{1}{\tau_B}$$

$$= \frac{2D_n}{W_B^2} \tag{8.8}$$

where D_n is the electron diffusion constant in the base, and W_B is the base width. Equation (8.8) shows that in bipolar transistors, ω_T is independent of the device size for a given process (given base width) at medium currents.

In an NMOS transistor, assuming the gate-to-source capacitance dominates, ω_T is

$$\omega_T = \frac{g_m}{C_{GS}} \tag{8.9}$$

For a saturated MOS transistor, $C_{GS} \approx \frac{2}{3}WLC_{OX}$. Thus, from Eqns. (8.5) and (8.9),

$$\omega_T \approx \frac{3}{L}\left[\frac{\mu_n I_D}{2WLC_{OX}}\right]^{\frac{1}{2}} \tag{8.10}$$

As in the case of intrinsic gain, f_T of an MOS transistor depends on many parameters including device sizes and drain current. In Eqn. (8.10), large I_{DS}, small size (small WL), and thick gate oxide (small C_{OX}) are preferable for high ω_T, which is exactly the opposite requirement for high gain from Eqn. (8.7). In bipolar transistors, from Eqn. (8.3) the intrinsic gain is constant, so that there is no trade-off between gain and frequency response. Therefore, bipolar transistors have better potential for high performance analog circuits where both high gain and high speed are required.

In many analog circuits, an amplifier has to drive a capacitive load. Switched capacitor filters and algorithmic A/D converters are good examples. The value of load capacitance is usually dictated by kT/C noise and ratio matching [8.15]. In such cases, the unity gain frequency (frequency at which voltage gain drops to unity) with the given capacitive loading is the most important frequency response parameter. An amplifier with high unity gain frequency can be clocked at a high rate, enabling high-frequency filtering or high-speed A/D conversion. For a common-emitter or common-source amplifier, the frequency response shown in Fig. 8.4 is obtained, assuming the pole introduced by the load capacitance C_L is the dominant pole. From the figure, the unity gain frequency can be found to be

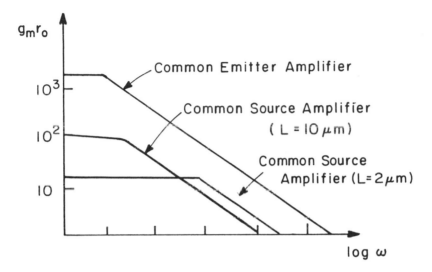

Fig. 8.4 Frequency response of common-emitter and common-source amplifiers. The frequency axis is of an arbitrary unit.

$$\omega_u = \frac{g_m}{C_T}$$ (8.11)

From Eqn. (8.11) it is obvious that a bipolar transistor amplifier has higher unity gain frequency, because bipolar transistors have much higher transconductance than MOS transistors at identical bias currents.

A meaningful comparison between the two devices can be made by examining the frequency response of a common-emitter or a common-source amplifier. The unity gain frequency is the upper boundary for the speed of the circuit, while the d-c gain (intrinsic gain) sets the upper boundary for the accuracy. In Fig. 8.4, frequency responses of a common-emitter amplifier and common-source amplifiers with two different gate lengths are plotted. Identical capacitive loading is assumed for both circuits. It is interesting to note that for the same current level the frequency response curves of MOS transistors are always inside that of bipolar transistors regardless of the gate lengths. This means that bipolar transistor circuits are capable of higher accuracy and speed.

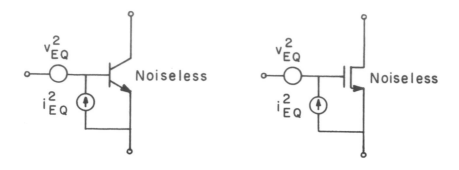

Fig. 8.5 Equivalent input noise generators

8.2.3 Noise

Noise in a device can be conveniently represented by equivalent input noise generators [8.16], as is shown in Fig. 8.5. The equivalent input voltage noise generator is in series with the input, while the equivalent input current generator is in parallel with the input. It is shown in Ref. 8.16 that in bipolar transistors

$$i_{EQ}^2 = 2q \ (I_B + K_1 \frac{I_B^a}{f} + \frac{I_C}{|\beta(j\omega)|^2} \ ) \, \Delta f \tag{8.12}$$

and

$$v_{EQ}^2 = 4kTr_b\Delta f + \frac{2qI_C\Delta f}{g_m^2} \tag{8.13}$$

In Eqns. (8.12) and (8.13), Δf is the noise bandwidth, and K_1 is the flicker noise coefficient. In MOS transistors

$$i_{EQ}^2 \approx 0 \tag{8.14}$$

$$v_{EQ}^2 = \frac{4kT\Delta f}{\frac{3}{2}g_m} + \frac{K\Delta f}{WLC_{OX}f} \tag{8.15}$$

Equivalent input current noise is small in MOS transistors at frequencies below the cut off frequency of the transistor. In Eqn. (8.15), K is the flicker noise coefficient.

Comparing noise equations (8.12) through (8.15), it can be seen that MOS transistors have much lower i_{EQ}^2 than bipolar transistors. On the other hand, v_{EQ}^2 is usually lower in bipolar transistors because g_m is much higher. When the device is driven by a low impedance source, i_{EQ} is shunted out, and only v_{EQ} is effective. In such cases, the circuit noise can be minimized by using bipolar transistors at the input. On the contrary, if the driving source has a high impedance, v_{EQ} is left open leaving only i_{EQ} effective. In such cases, an MOS input will minimize circuit noise. Therefore, to minimize circuit noise, proper devices should be selected at the input, depending on the impedence of the circuit driving the input. In some applications, however, other factors such as input leakage current (base current) dictate the choice of input devices.

8.2.4 Offset Voltage

An important accuracy parameter in devices is the offset voltage of a differential amplifier caused by mismatches. In a resistively loaded bipolar differential amplifier (differential amplifier with collector resistors), the input offset voltage is shown to be [8.16]

$$V_{OS} = \frac{kT}{q}\left[-\frac{\Delta R}{R} - \frac{\Delta A_E}{A_E} - \frac{\Delta Q_B}{Q_B} \right] \qquad (8.16)$$

where R is the load resistor, A_E is the emitter area, and Q_B is the base Gummel number. Δ represents the mismatch between the parameters. For the purpose of comparison, consider a resistively loaded MOS differential amplifier. From a similar analysis to that of a bipolar differential pair, MOS differential amplifier offset is obtained [8.16].

$$V_{OS} = \Delta V_T + \frac{V_{GS} - V_T}{2}\left[-\frac{\Delta R}{R} - \frac{\Delta(W/L)}{W/L} \right] \qquad (8.17)$$

The mismatch terms inside the brackets in both Eqns. (8.16) and (8.17) are caused by process variations and photolithographic resolution, and they are assumed to be of a comparable magnitude. In case of MOS offset voltage, the multiplier term $(V_{GS} - V_T)/2$ is on the order of 500 mV, while for bipolar offset, kT/q is only 25 mV. This gives more than an order of magnitude smaller offset voltage in a bipolar differential amplifier. Also, for MOS offset, there is a threshold voltage mismatch term ΔV_T, which is absent in bipolar offset. In an MOS differential amplifier, offset voltages of 5-10 mV are common, while bipolar differential amplifiers typically have offset voltage under 1 mV. It should be noted that the MOS offset voltage can be reduced by decreasing the current in the device, which gives smaller $V_{GS} - V_T$. This contradicts, however, the high-speed requirements discussed in Section 8.2.2.

8.2.5 Other Parameters

There are many other differences between bipolar and MOS transistors, which can be significant depending on the application. For example, In an *n*-well BiCMOS process, NMOS transistors suffer the back-gate effect, while *npn* transistors do not. Typical V_{BE} is about 0.6 V, which may be preferred to a V_{GS} of typically 1 V, for low voltage operations. For capacitor switching, MOS transistors are much more ideal because the voltage drop between the drain and the source is zero in steady-state. In contrast, bipolar transistors have a few tenth of volts between the collector and the emitter when it is saturated [8.17]. Also, MOS transistors can be near ideal current switches compared to bipolar transistors which introduce error due to finite β. Such differences should be carefully examined before determining which type of devices are to be used in a particular part of the circuit.

8.3 Analog Subcircuits

In this section, analog subcircuits such as current mirrors and basic amplifier stages are discussed. Current mirrors are frequently used in analog circuits as biasing circuits, active loads, or bit current cells in A/D and D/A converters. Different types of bipolar, MOS, and BiCMOS current mirrors are discussed in Section 8.3.1. In Section 8.3.2, a number of different basic amplifiers using both MOS and bipolar transistors are described.

8.3.1 Current Mirrors

For a current mirror to behave like a current source, the ouput resistance of the mirror must be high. For this reason, the output resistance is the most important parameter for current mirrors. In simple current mirrors (Fig. 8.6 and 8.7), the output resistance of the mirror is the same as the output resistance r_o of the transistors at I_O. Thus for the bipolar current mirror,

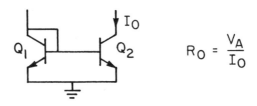

Fig. 8.6 Simple bipolar current mirror.

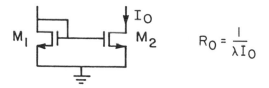

Fig. 8.7 Simple MOS current mirror.

$$R_o = r_{o,q2} = \frac{V_A}{I_O} \tag{8.18}$$

and for the MOS current mirror,

$$R_o = r_{o,m2} = \frac{1}{\lambda I_O} \tag{8.19}$$

A bipolar implementation has higher output resistance due to higher Early Voltage. Although it is possible to achieve higher output resistance in a MOS current mirror by using long channel transistors, this will lower the parasitic pole frequency of the mirror. Also, unless the gate width is scaled proportionally, the gate-to-source voltage may have to be excessively large, and the output swing is limited. This is because the saturation voltage of an MOS transistor increases with the gate length [8.16].

$$V_{DS,SAT} \approx V_{GS} - V_T = \left[\frac{2I_D L}{\mu_n C_{OX} W} \right]^{\frac{1}{2}} \tag{8.20}$$

Often, higher output resistance is desired than is available from a simple current mirror. Cascode and Wilson current mirrors are popular for high output resistance applications. In modern processes where supply voltages are low, cascode current sources are preferable, because it is possible to improve the output swing (range of output voltage for which the output resistance remains high). Cascode current mirrors with improved output swing are discussed in detail elsewhere [8.16, 8.18], and will not be discussed further here.

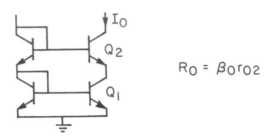

$$R_O = \beta_O r_{O2}$$

Fig 8.8 Bipolar cascode current mirror.

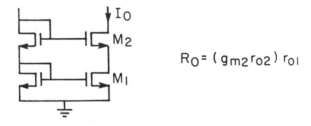

$$R_O = (g_{m2} r_{o2}) r_{o1}$$

Fig 8.9 MOS cascode current mirror.

In the bipolar cascode current mirror shown in Fig. 8.8, the output resistance can be found from the small-signal equivalent circuit. Again, neglecting r_μ, r_{ex} and r_b, the output resistance of the bipolar cascode current mirror is shown to be [8.16]

$$R_o = \beta_o r_{o2} \tag{8.21}$$

where r_{o2} is the output resistance of Q_2. In the MOS cascode current mirror of Fig. 8.9, the output resistance is [8.16]

$$R_o = (g_{m2} r_{o2}) r_{o1} \tag{8.22}$$

where g_{m2} is the transconductance of M_1, and r_{o1} and r_{o2} are the output resistances of M_1 and M_2, respectively.

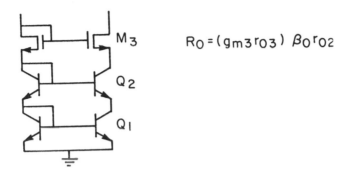

$$R_O = (g_{m3}r_{03}) \beta_0 r_{02}$$

Fig. 8.10 BiCMOS double cascode amplifier.

In the bipolar cascode current mirror, the improvement on the output resistance from the simple current mirror is a factor of β while in the MOS cascode current mirror it is a factor of $g_m r_o$. In bipolar transistors, finite β limits the improvement below the intrinsic gain $g_m r_o$. In normal current ranges, β is usually larger than $g_m r_o$ of a short-channel MOS transistor. Thus, combined with initially higher output resistance of a bipolar current mirror, bipolar cascode current mirrors have higher output resistance than MOS cascode current mirrors in general. However, it should be noted that for an MOS transistor at very low current densities, especially in the subthreshold region, $g_m r_o$ can be very large, and can exceed β_0 of the bipolar transistor. Therefore, higher output resistance can be obtained from an MOS current mirror at the cost of the frequency response.

Yet higher output resistance can be obtained by double cascode current mirrors. The bipolar double cascode mirror does not provide any improvement from the cascode mirror, and should not be used. This is because of the finite current gain limitation in bipolar transistors. A BiCMOS double cascode shown in Fig. 8.10 further improves the output resistance of a bipolar cascode by a factor of the intrinsic gain of the MOS transistor. In such a high output resistance circuit, stray conductance can limit the actual output resistance, and care must be taken to minimize the parasitic conductance at the output node. The output resistance of the BiCMOS double cascode mirror is shown to be

$$R_o = \beta_o(g_{m3}r_{o3})\, r_{o2} \qquad (8.23)$$

8.3.2 Basic Amplifier Circuits

A bipolar common-emitter amplifier with an active load provides high voltage gain, which can be calculated from the small-signal equivalent circuit shown in Fig. 8.11 [8.16].

$$A_v = \frac{r_\pi}{\eta(r_\pi + r_b)} \tag{8.24}$$

where

$$\eta = \frac{kT}{qV_A} \tag{8.25}$$

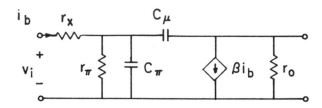

Fig. 8.11 Small-signal equivalent circuit for a common-emitter amplifier.

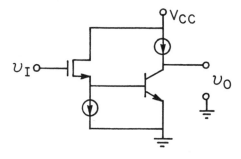

Fig. 8.12 Source-follower/common-emitter combination.

For the typical Early Voltage of 50 V, and assuming $r_b \ll r_\pi$, a voltage gain of 2,000 is obtained at room temperature. The input resistance of a common-emitter amplifier is $r_b + r_\pi$ which is on the order of a few kilo ohms. The obvious thing which can be done to increase the input resistance is to use a source follower before the common-emitter amplifier as shown in Fig. 8.12. A small-signal analysis on this circuit shows that the voltage gain is

$$A_v = \frac{r_\pi}{\eta\left(\dfrac{1}{g_{m1}} + r_b + r_\pi\right)} \tag{8.26}$$

Compared with a common-emitter amplifier, some degradation of frequency response is to be expected. The parasitic pole due to C_π can be approximately calculated by the open-circuit time constant method [8.19].

$$\omega_p = \frac{1}{\left(\dfrac{1}{g_{m1}} + r_b\right) \| \, r_\pi \, C_\pi} \tag{8.27}$$

Since g_{m1} is relatively small and C_π is usually large, this parasitic pole can affect the frequency response significantly, especially when the circuit is connected in feedback as is the case for operational amplifiers.

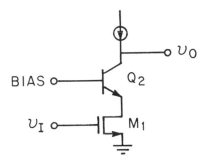

$R_i = \infty$

$A_\upsilon = g_{m1}(\beta_0 r_{02})$

Fig. 8.13 BiCMOS cascode amplifier.

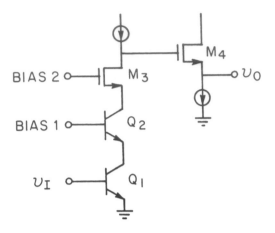

Fig. 8.14 BiCMOS double cascode amplifier.

Another way of achieving high gain without the above problem is to use a cascode configuration. In Fig. 8.13, an MOS common-source amplifier is followed by a bipolar common-base stage. If this circuit is loaded with a high output resistance current source, the voltage gain can be quite large, with infinite input resistance. Unlike the source-follower/common-emitter combination, the parasitic pole added due to cascoding is at a very high frequency. This is because the parasitic pole is determined by the transconductance of the bipolar transistor and the capacitance at the emitter of the bipolar transistor. The parasitic pole frequency should be a fraction of the f_T of the bipolar transistor. The voltage gain of this cascode amplifier is the product between the transconductance g_{m1} and the output resistance of the cascode. Thus,

$$A_v = g_{m1}(\beta_o r_{o2}) \qquad\qquad (8.28)$$

Extremely high voltage gain can be obtained from a single voltage gain stage, using a bipolar/MOS double cascode amplifier of Fig. 8.14 [8.20]. Assuming an ideal current source loading, the voltage gain of this amplifier can be approximated by the product between the transconductance of the input bipolar transistor and the output resistance of the circuit. The output resistance of this circuit should be extremely high, since the circuit is identical to the right half of the bipolar-MOS double cascode current mirror discussed in Section 8.3.1. In practice, the output resistance will be limited by the output resistance of the current source load. A PMOS double cascode current mirror with relatively long gate lengths can be used as the load. From the output resistance of the double cascode, the voltage gain A_v can be found.

$$A_v = g_{m1} \cdot R_o$$

$$= \beta_o (g_{m1} r_{o1})\, (g_{m2} r_{o2}) \tag{8.29}$$

$$= \frac{\beta_o}{\eta} (g_{m1} r_{o1}) \tag{8.30}$$

Care must be taken to avoid stray conductance at the output due to extremely high output resistance, and it is necessary to employ a source follower if any resistive load is to be driven.

8.4 Operational Amplifiers

In this section, BiCMOS operational amplifier design techniques are discussed. To take full advantage of BiCMOS processes in operational amplifier design, it is important to understand the limitations in bipolar or CMOS operational amplifiers. In Section 8.4.1, basic two-stage operational amplifiers in bipolar and MOS technologies are discussed, and in Section 8.4.2, frequency responses of two-stage operational amplifiers are analyzed. In Section 8.4.3, BiCMOS operational amplifier design techniques are described, and in Section 8.4.4, slew rate and noise of BiCMOS operational amplifiers are considered.

8.4.1 Basic Two Stage Operational Amplifier

For general purpose operational amplifiers, large DC voltage gain is required. Two stage operational amplifiers are the most popular among general purpose operational amplifiers because much larger gain can be obtained compared to single-stage circuits such as a folded-cascode design. Although three or more gain stages would provide more voltage gain, frequency compensation of such circuits is very difficult, if not impossible. In this section, basic two-stage operational amplifiers are described.

Since both high gain and high frequency characteristics are desired in operational amplifiers, bipolar transistors are better devices for this purpose, as was discussed in Section 8.2. A simplified all-bipolar operational amplifier circuit is shown in Fig. 8.15. This circuit can be considered as a simplified schematic of the popular µA 741. It should be noted that in the 741, an *npn-pnp* quad input is used instead of a pair of *pnp* transistors, and emitter followers are added at the output and between the first and the second stages. The problem with all-bipolar design like the one in Fig. 8.15 is that the input resistance tends to be small (on the order of a mega ohms, at most), and input bias currents are large (on the order of hundreds of nano amperes). Also, very low f_T of lateral *pnp* transistors limits the useful frequency range significantly.

On the other hand, an all-MOS operational amplifier shown in Fig.8.16 has compromised gain and bandwidth due to the nature of MOS transistors as discussed in Section 8.2. Often, much longer than minimum gate lengths are used to achieve acceptable gain at the expense of bandwidth. Due to large flicker noise in MOS transistors, long gate lengths are also needed in the input stage as discussed later in Section 8.4.4. Higher gain can be obtained by cascoding the first stage as shown in Fig. 8.17. Although cascoding the second stage is possible, it reduces the ouput swing.

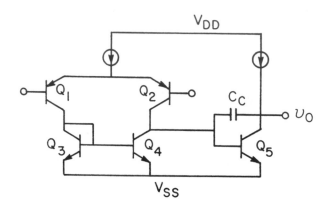

Fig 8.15 All-bipolar operational amplifier.

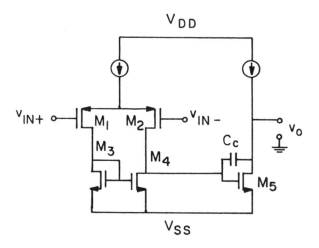

Fig. 8.16 All-MOS operational amplifier.

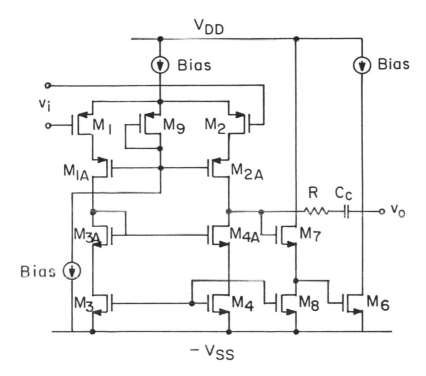

Fig. 8.17 Cascoded CMOS operational amplifier.

8.4.2 Frequency Response of Two Stage Op Amp

In this section, frequency response and compensation of two stage operational amplifiers is discussed. A small-signal equivalent circuit of a basic two stage operational amplifier is shown in Fig. 8.18. In this circuit, G_{m1}, R_{o1}, G_{m2}, and R_{o2} are the transconductances and the output resistances of the first and second stage, respectively. C_1 is the total parasitic capacitance at the output of the first stage, and C_2 is the sum of the parasitic capacitance and any load capacitance present at the output.

For such an amplifier, the open loop DC gain is

$$A_v = G_{m1}R_{o1}G_{m2}R_{o2} \tag{8.31}$$

With the pole-splitting compensation by C_c, the dominant pole is [8.16]

$$|p_1| \approx \frac{1}{R_{o1}C_cG_{m2}R_{o2}} \tag{8.32}$$

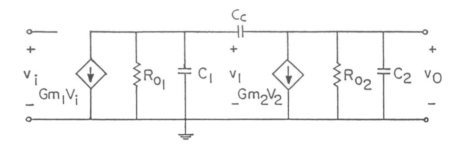

Fig. 8.18 Small-signal equivalent circuit for a two-stage operational amplifier.

and the non-dominant pole p_2 [8.16]

$$|p_2| \approx \frac{G_{m2}}{C_2} \qquad (8.33)$$

From Eqns. (8.31) and (8.32), the unity gain bandwidth is found to be

$$\omega_1 \approx \frac{G_{m1}}{C_c} \qquad (8.34)$$

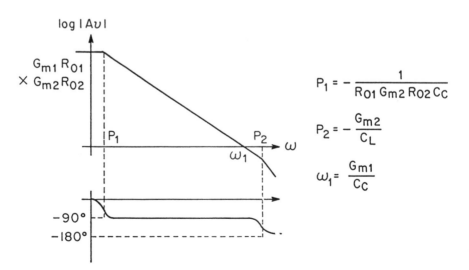

$$P_1 = -\frac{1}{R_{O1} \, G_{m2} \, R_{O2} \, C_c}$$

$$P_2 = -\frac{G_{m2}}{C_L}$$

$$\omega_1 = \frac{G_{m1}}{C_c}$$

Fig. 8.19 Bode plot of compensated operational amplifier.

If $\omega_1 = |p_2|$, the phase margin would be 45° in the unity gain feedback configuration because p_2 is where the phase of the loop transfer function becomes 135°. Generally, a larger phase margin is desirable, in order to reduce peaking in the frequency response and overshoot in the step response. Therefore, the value of the compensation capacitor C_c must be selected so that ω_1 is below p_2 (see Fig. 8.19). For this reason, high frequency operational amplifiers must be designed for high p_2, or high G_{m2}, for given load capacitance.

8.4.3 BiCMOS Operational Amplifiers

One obvious approach to solve the input resistance/bias current problem of the all-bipolar design, although not a good one, is to buffer the input with a pair of source followers, as shown in Fig. 8.20. Although this circuit offers infinite input resistance and zero input bias current, the frequency response is adversely affected. This is because the low transconductance of the MOS source follower is driving the large C_π of the bipolar transistor. A brute-force solution to this problem is to increase the transconductance of the input MOS transistors by increasing their size and bias current. This is hardly economical, since the transconductance is only a square root function of the MOS transistor size and bias current.

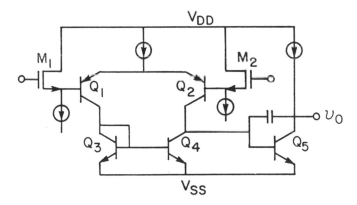

Fig. 8.20 Source-follower buffered bipolar operational amplifier.

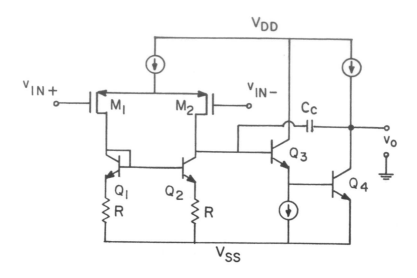

Fig. 8.21 BiCMOS operational amplifier.

An alternative design is shown in Fig. 8.21, in which a PMOS source-coupled pair is used in the input. In this design, high p_2 is obtained from the high transconductance of the bipolar second stage, while the PMOS input stage provides infinite input resistance. The input resistance of the second stage amplifier is $r_b + r_\pi$ (on the order of a few kilo ohms), which is much smaller than the output resistance of the first stage. To reduce this loading of the second stage amplifier on the first stage, an emitter-follower is added. The frequency response of this circuit should be better than the source-follower buffered bipolar design due to the absence of the potentially low frequency parasitic pole discussed above. The emitter degeneration resistors in the current mirror are added to reduce the noise contribution from the bipolar current mirror. As will be discussed in the next section, the PMOS input differential pair improves slew rate compared with a bipolar differential pair.

Another operational amplifier design that can benefit from a BiCMOS technology is a folded-cascode circuit. A CMOS folded-cascode amplifier is shown in Fig. 8.22. In a folded-cascode amplifier, frequency compensation is achieved by the load capacitance C_L [8.21]. For such an amplifier, the DC gain is the product between the transconductance of the input differential pair and the output resistance.

$$A_v = G_{m1}R_o \tag{8.35}$$

where G_{m1} is the transconductance of the input differential pair, and R_o is the output resistance at the output node. The dominant pole is due to R_o and C_L:

$$|p_1| \approx \frac{1}{R_o C_L} \tag{8.36}$$

There are parasitic poles associated with the common-gate stages and the current mirror. It is very likely that the lowest frequency parasitic pole is due to the common-gate stage. It is because the input transistors are typically very large, and contribute correspondingly large parasitic capacitance at the input of the common-gate stage. In this case, the non-dominant pole is at

$$|p_2| \approx \frac{g_{m3-4}}{C_p} \tag{8.37}$$

. where g_{m3-4} is the transconductance of the common-gate transistors (M3 and M4), and C_p is the total (parasitic) capacitance at the source of the common-gate transistor. From Eqns. (8.35) and (8.36), the unity-gain frequency can be shown to be

$$\omega_1 = \frac{G_{m1}}{C_L} \tag{8.38}$$

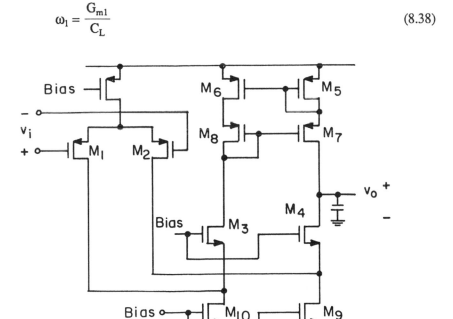

Fig. 8.22 CMOS folded-cascode operational amplifier.

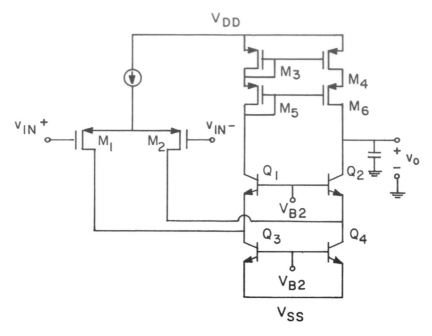

Fig. 8.23 BiCMOS folded-cascode operational amplifier.

For fixed load capacitance C_L, the input transconductance G_m should be made small enough so that $\omega_1 < |p_2|$, to ensure a phase margin greater than 45°.

In Fig. 8.23, the common-gate stages are replaced with common-base stages so that $|p_2|$, and consequently ω_1, can be made much higher due to higher transconductance from bipolar transistors.

When it is necessary to drive a resistive load, a low resistance output buffer is needed. A bipolar emitter follower is an ideal circuit for this purpose, because of its high input resistance and low output resistance. When a large capacitive load is present at the output, a complementary follower circuit as shown in Fig. 8.24 can be added. Rather than a slow lateral *pnp* transistor, a short channel PMOS transistor is employed. To minimize the reduction of output swing due to the back-gate effect, the *n*-well of the PMOS transistor must be connected to the source.

8.4.4 Slew Rate and Noise Considerations

For a two stage operational amplifier like the one in Fig. 8.15, slew rate is determined by the input stage bias current I_x and the compensation capacitor C_c, because the second stage acts as an integrator, which integrates the output current from the first stage.

$$SR = \frac{I_x}{C_c} \qquad (8.39)$$

Although from Eqn. (8.39) it appears that slew rate can be improved by increasing I_x or by reducing C_c, in an all-bipolar design or source follower buffered bipolar operational amplifier, slew rate is fixed by other factors. From Eqn. (8.34),

$$C_c = \frac{G_{m1}}{\omega_1} \qquad (8.40)$$

Now, Eqn. (8.40) becomes

$$SR = \frac{I_x}{G_{m1}} \omega_1 \qquad (8.41)$$

Since ω_1 must be made smaller than $|p_2|$ for phase margin greater than 45°,

$$SR < \frac{I_x}{G_{m1}} |p_2| \qquad (8.42)$$

The input stage current I_x is shared between two transistors and the collector current in each of the input transistors is $I_x/2$. Thus,

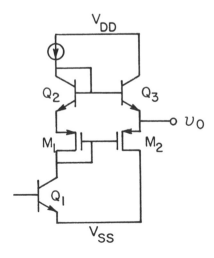

Fig. 8.24 Complementary BiCMOS output buffer.

$$G_{m1} = \frac{qI_x}{2kT} \tag{8.43}$$

or

$$\frac{I_x}{G_{m1}} = \frac{2kT}{q} \tag{8.44}$$

From Eqns. (8.42) and (8.44), maximum slew rate SR_{max} is

$$SR_{max} = \frac{2kT}{q} |p_2| \tag{8.45}$$

Equation (8.45) shows that the upper bound of slew rate is determined by $|p_2|$. One way of improving slew rate is to increase $|p_2|$. Slew rate can also be increased by improving the ratio I_x/G_{m1}. For a bipolar differential pair, this can be achieved by inserting emitter degeneration resistors. If MOS differential pair is used as in Fig. 8.21, from Eqn. (8.5),

$$\frac{I_x}{G_{m1}} = (V_{GS} - V_T) \tag{8.46}$$

From Eqns. (8.42) and (8.46),

$$SR_{max} = (V_{GS} - V_T)|p_2| \tag{8.47}$$

Comparing Eqns. (8.45) and (8.47), $(V_{GS} - V_T)$ is usually more than an order of magnitude larger than kT/q. For this reason, an MOS input operational amplifier has potentially much higher slew rate than bipolar input counterparts. It is obvious that the circuit shown in Fig. 8.21 has excellent slew rate, because it has MOS input transistors, and also because $|p_2|$ is high.

For a two-stage operational amplifier, input equivalent noise is shown to be [8.16]:

$$v_{EQ,total}^2 = 2v_{EQ,input}^2 \left[1 + \frac{g_{m,mirror}^2}{g_{m,input}^2} \right] \tag{8.48}$$

From Eqn. (8.48) it is clear that the transconductance of the current mirror must be minimized to reduce the second term, which is the noise contribution from the current mirror. If MOS input transistors and a bipolar current mirror is used as shown in Fig. 8.21, without the emitter degeneration resistors, the current mirror has much larger transconductance than that of input transistors giving poor noise performance. To improve noise in such a case, emitter degeneration resistors can be employed in the current mirror, to reduce the effective transconductance as shown in Fig. 8.21. The emitter degeneration resistors reduce the transconductance of the current mirror by a factor of $(1 + g_m R_E)$ [8.16].

8.5 BiCMOS Analog Subsystems

A wide variety of analog subsystems such as A/D and D/A converters, log/antilog circuits, translinear circuits, and filter circuits can be implemented in BiCMOS technologies. Because high-performance analog BiCMOS technologies are just emerging, this is a fertile field for circuit design, although at this time there are only a few design examples. In this section, some circuit design ideas are investigated for the application of high-performance BiCMOS technologies in analog subsystems.

8.5.1 A/D and D/A Converters

Among many different types of A/D converters, flash A/D converters would probably benefit most from BiCMOS technologies. For an N bit flash A/D converter, $2^N - 1$ voltage comparators are needed. Since the conversion speed

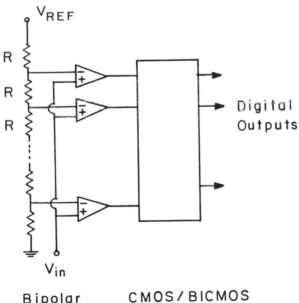

Bipolar   CMOS/BICMOS
Comparators   Encoding Logic

Fig. 8.25 Block diagram of a BiCMOS flash A/D converter.

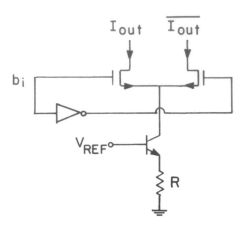

Fig. 8.26 BiCMOS current switching D/A converter cell.

speed is limited mostly by the comparator delay, the comparator must be optimized for high speed. However, due to the large number of comparators in a flash A/D converter, they must also be small and low-power. Since bipolar transistors offer higher transconductance and speed for given power and area, it is logical to design the comparators with bipolar transistors. On the other hand, the encoding logic can be built in CMOS, taking advantage of the high packing density and low power of CMOS logic circuits. Pipelined encoding logic can overcome the slower speed of CMOS logics compared with bipolar ECL logic circuits. A block diagram of such a BiCMOS flash A/D converter is shown in Fig. 8.25.

In high resolution current switching D/A converters, the base current error in the steering switches can pose serious problems in accuracy and temperature stability. Although Darlington-connected switches can reduce the base current error, switching speed is reduced. Instead, MOS switches can be used to eliminate the error. An example of a BiCMOS bit current cell employing MOS steering switches and bipolar current sources is shown in Fig. 8.26.

Although it is possible to implement a similar current bit cell design in all-MOS, so that it can be built in a CMOS technology, the BiCMOS design will give a better accuracy. Bipolar current cells have better current matching, determined mainly by the matching of emitter resistors. If the voltage drop across the emitter degeneration resistor is substantially larger than the thermal voltage (25 mV at room temperature), V_{BE} mismatch becomes insignificant [8.16]. On the other hand, in MOS current cells, source resistors are far less effective than emitter resistors. A voltage drop much larger than $V_{GS} - V_T$ across the source resistor is required to reduce the effect of V_T and W/L mismatches in MOS transistors. Since $V_{GS} - V_T$ is already on the order of a volt, as much as 10 V must be dropped across the source resistor to reduce the V_T and W/L mismatches by a factor of 10. This is certainly impractical in most MOS technologies with the maximum supply voltage of 5-10 V.

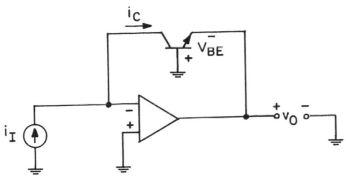

Fig. 8.27 Simple log circuit.

Also, bipolar current sources are quieter because there is much less 1/f noise in bipolar transistors. In bipolar current sources with emitter degeneration, transistor noise is usually smaller than noise from the degeneration resistors. In this case, the output current noise power is inversely proportional to the emitter resistor [8.16]. Therefore, to reduce noise, large emitter degeneration resistors are desired. A trade-off must be made to optimize noise, resistor area, and the voltage drop across the resistor.

The performance of other types of A/D converters such as charge-redistribution A/D converters, cyclic converters, or pipeline converters [8.1, 8.7, 8.22-8.23] can be improved in BiCMOS technologies by utilizing bipolar transistors in the operational amplifiers and voltage comparators, while MOS transistors are used as switches and in the logic circuits. An appreciable savings in die area can be made by redesigning the successive approximation register alone in CMOS in an all-bipolar A/D converter.

8.5.2 Log/Antilog Circuits

By utilizing the excellent conformity of the collector current to an exponential characteristic, log or antilog circuits can be implemented. A simple example is shown in Fig. 8.27. The output voltage of the circuit is

$$v_O = \frac{kT}{q} \ln \frac{i_I}{I_s} \tag{8.49}$$

where I_s is the collector saturation current of the transistor.

The drawback of this simple circuit is that the output voltage depends on I_s, which is a highly process and temperature dependent parameter. The other problem is the input bias current of the operational amplifier, which takes some of the input current away, giving inaccuracies in the resulting output.

In the circuit shown in Fig. 8.28, the output voltage dependence on I_s is removed by an additional transistor and an operational amplifier. MOS input operational amplifiers are used to eliminate the error due to the operational

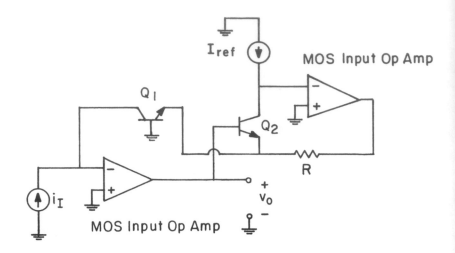

Fig. 8.28 Precision log circuit.

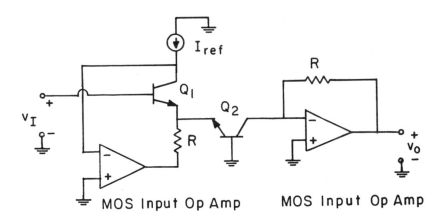

Fig. 8.29 Precision antilog circuit.

amplifier input bias current. Although an MOS input operational amplifier has larger offset voltage, it has little effect on accuracy. The input offset voltage causes the collector voltage mismatch in Q_1 and Q_2. The ratio mismatch of collector currents due to the collector current mismatch is

$$\frac{\Delta I_C}{I_C} \approx \frac{\Delta V_{CE}}{V_A} \tag{8.50}$$

where ΔV_{CE} is the collector voltage mismatch and V_A is the Early Voltage of the transistor.

With ΔV_{CE} of 10 mV and V_A of 30 V, the ratio error is only 3.3×10^{-4}, which is much less than the typical ratio error caused by I_s mismatch (on the order of 1 %). The output voltage of the circuit is

$$v_O = v_{BE2} - v_{BE1}$$

$$= \frac{kT}{q} \ln \frac{i_I}{I_{ref}} \qquad (8.51)$$

Notice that I_s dependence is removed in (8.51). It is possible to modify this circuit for a voltage input by connecting a resistor between the input voltage and the inverting input of the operational amplifier to generate $i_I = v_I/R$. In this case, however, the non-zero offset voltage of the operational amplifier limits the accuracy so that $i_I = (v_I - V_{OS})/R$. If high accuracy is desired in the voltage input mode, offset cancellation techniques such as double-correlated sampling or chopper stabilization should be employed.

In a similar manner, an antilog circuit can be built. A schematic diagram of an antilog circuit is shown in Fig. 8.29. The output voltage can be shown to be

$$v_O = RI_{ref} \exp \frac{qv_I}{kT} \qquad (8.52)$$

8.5.3 Translinear Circuits

Translinear circuits are the type of circuits that perform controlled non-linear operations generated from the exponential current-voltage characteristic of bipolar transistors. An example is the analog multiplier shown in Fig. 8.30. In this circuit, i_X and i_Y are the input currents, and the output is the differential collector current $i_2 - i_1$. Applying Kirchhoff's Voltage Law around Q_1 through Q_4,

$$v_{BE3} - v_{BE4} = v_{BE2} - v_{BE1} \qquad (8.53)$$

Neglecting base currents, and assuming that the transistors are ideally matched (identical I_s), Eqn. 8.53 becomes

$$\frac{i_{C3}}{i_{C4}} = \frac{i_{C2}}{i_{C1}} \qquad (8.54)$$

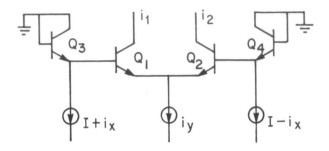

Fig. 8.30 Translinear analog multiplier.

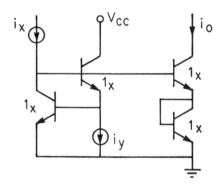

Fig. 8.31 Translinear circuit example; $i_O = \sqrt{i_X \cdot i_Y}$.

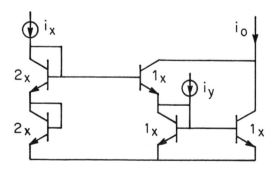

Fig. 8.32 Translinear circuit example; $i_O = \sqrt{i_X^2 + i_Y^2}$.

or

$$\frac{I + i_X}{I - i_X} = \frac{i_2}{i_1} \tag{8.55}$$

Also,

$$i_1 + i_2 = i_Y \tag{8.56}$$

From Eqns. (8.55) and (8.56),

$$i_O = i_2 - i_1$$

$$= \frac{i_X \cdot i_Y}{I} \tag{8.57}$$

Other examples of translinear circuits is shown in Fig. 8.31 and Fig. 8.32. For the circuit in Fig. 8.31,

$$i_O = \sqrt{i_X \cdot i_Y} \tag{8.58}$$

and for the circuit in Fig. 8.32,

$$i_O = \sqrt{i_X^2 + i_Y^2} \tag{8.59}$$

The strength of translinear circuits is that controlled non-linear functions can be generated from very simple circuitry, and that the bandwidth of the circuit is high. In the current mode operation, bandwidth exceeding 10 MHz can be obtained [8.24].

Good quality bipolar transistors are essential in translinear circuits for both speed and accuracy. Lateral bipolar transistors are generally avoided due to their low α and f_T.

There are many interesting possibilities of applications if translinear circuits are combined with MOS circuits, by using a BiCMOS technology. For example, in early vision processing systems, an array of a large number of translinear circuits can be built on a single silicon chip which handle the necessary processing much more efficiently and rapidly than the digital counterpart. Compact and fast A/D converters such as CMOS/BiCMOS cyclic or pipelined converters [8.22, 8.23] can be built on the same chip for rapid conversion of processed signals for communication with a digital computer which will handle middle and late vision processing.

8.5.4 Other Analog and Analog/Digital Circuits

Since a BiCMOS technology provides the advantages of both bipolar and MOS technologies, there are, and will be, numerous applications which are not covered in this chapter. Although many analog circuits can still be built in either all-bipolar or all-MOS technologies, BiCMOS technologies render the circuit designers a tremendous flexibility. Provided the designers are adequately trained in BiCMOS circuits, the design time can be shorter with BiCMOS technologies due to the flexibility of design. There is little to dispute that BiCMOS designs will be of higher performance and lower power or area.

Phase-locked loops for high-speed data communication systems would be another circuit for which a BiCMOS technology can be useful. Bipolar voltage-controlled oscillators can be built for high speed, good linearity, and low temperature and supply sensitivity [8.25]. In an all-MOS design, due to the inherent limitation caused by the lower transconductance in MOS transistors, some performance parameters have to be traded with speed [8.26]. On the other hand, the charge-pump type phase detector/loop filter can be implemented more easily in CMOS technologies. With a BiCMOS technology, it is possible to choose the best topologies for the VCO's and the phase detectors, and also possible to integrate complex data processing digital circuits following the phase-locked loop on the same die.

In SRAM's, the analog part of the circuit is the sense amplifier. Lower power/higher speed can be achieved from bipolar sense amplifiers, while bit cells are implemented in CMOS for lower power and high packing density [8.27-8.29]. Quite obviously, the same strategy can be applied in any analog/digital mixed systems; bipolar analog for high performance and CMOS digital for low power/high density. Another step forward from this is to utilize true BiCMOS circuits for both the analog and the digital portion to enhance the performance. This would seem to be a difficult task from the designer's point of view. However, by applying some of the techniques discussed in this chapter and preceding chapters and creative thinking, it is hoped that true BiCMOS analog/digital design will become a routine task.

REFERENCES

8.1] J.L. McCreary and P.R. Gray, "All-MOS charge redistribution analog-to-digital conversion techniques-Part I," IEEE J. Solid-State Circuits, vol. SC-10, pp. 371-378, Dec. 1975.

8.2] I.A. Young, D.A. Hodges, and P.R. Gray, "Analog NMOS sampled-data recursive filter," in ISSCC Dig. Tech. Papers, pp. 156-157, Philadelphia, PA, Feb. 1977.

8.3] Y.P Tsividis and P.R. Gray, "An integrated NMOS operational amplifier with internal compensation," IEEE J. Solid-State Circuits, vol. SC-11, pp/748-753, Dec. 1976.

8.4] D. Senderowicz, S.F. Dreyer, J.H. Huggins, C.F Rahim, and C.A. Laber, "A family of differential NMOS analog circuits for a PCM CODEC filter chip," IEEE J. Solid-State Circuits, vol. SC-17, pp. 1014-1023, Dec. 1982.

8.5] B.J. White, G.M. Jacobs, and G.F. Landsburg, "A monolithic dual tone multifrequency receiver," IEEE J. Solid-State Circuits, vol. SC-14, pp. 991-997, Dec. 1979.

8.6] Y.A. Haque, R Gregorian, R.W. Blasco, R.A. Mao, and W. E. Nicholson, Jr., "A two chip PCM voice CODEC with filters," IEEE J. Solid-State Circuits, vol. SC-14, pp. 961-969, Dec. 1979.

8.7] H.-S. Lee, D.A. Hodges, and P.R. Gray, "A self-calibrating 15 bit CMOS A/D converter," IEEE J. Solid-State Circuits, vol SC-19, pp. 813-819, Dec. 1984.

8.8] B. K. Ahuja and W.M. Baxter, "A programmable dual channel interface processor," in ISSCC Dig. Tech. Papers, pp. 232-233, San Fransisco, CA, Feb. 1984.

8.9] M. Tadauchi, N. Hamada, K. Sato, K. Yasunari, Y. Nagayama, K. Nagai, N. Suemori, and T. Kubo, "A CMOS facsimile video signal processor," IEEE J. Solid-State Circuits, vol. SC-20, pp. 1179-1184, Dec, 1985.

8.10] K.K. O, H.-S. Lee, R. Reif, and W. Frank, "A 2 μm BiCMOS process utilizing selective epitaxy," IEEE Electron Device Letters, vol. 9, pp. 567-569, Nov. 1988.

8.11] T. Yamaguchi, Y. Wakui, K. Inayoshi, C. Tsuchiya, and M. Tokuriki, "20V BiCMOS technology with polysilicon emitter structure," in Extended Abstracts, Electrochem. Soc. Meeting (Spring), pp. 419-420, Philadelphia, PA, May, 1987.

8.12] T. Ikeda, A. Watanabe, Y. Nishio, I. Masuda, N. Tamba, M. Odaka, and K. Ogiue, "High-speed BiCMOS technology with a buried twin well structure, IEEE Trans. Electron Devices, vol. ED-34, pp. 1304-1310, June, 1987.

8.13] T.L. Tewksbury, H.-S. Lee, and G.A. Miller, "The effects of oxide traps on the large-signal transient response of MOS circuits," to appear in IEEE J. Solid-State Circuits.

8.14] J.L. McCreary, "Matching properties, and voltage and temperature dependence of MOS capacitors," IEEE J. Solid-State Circuits, vol. SC-16, pp. 608-616, Dec. 1981.

8.15] R. Castello and P.R. Gray, "Performance limitations in switched-capacitor filters," IEEE Trans. Circuits Syst. vol. CAS-32, pp. 865-876, Sept. 1985.

8.16] P.R. Gray and R.G Meyer, Analysis and Design of Analog Integrated Circuits, 2nd ed. New York, Wiley, 1984.

8.17] R.S. Muller and T.I. Kamins, Device Electronics for Integrated Circuits, 2nd ed. New York, Wiley, 1986.

8.18] R. Gregorian and G.C. Temes, Analog MOS Integrated Circuits for Signal Processing, New York, Wiley, 1986.

8.19] P.E. Gray and C.L. Searle, Electronic Principles, Physics, Models, and Circuits, New York, Wiley, 1969.

8.20] J.K. Roberge, Operational Amplifiers: Theory and Practice, New York, Wiley, 1975.

8.21] T.C. Choi, R.T. Kaneshiro, R.W. Brodersen, P.R. Gray, W.B. Jett, and M. Wilcox, "High-frequency CMOS switched-capacitor filters for communications application," IEEE J. Solid-State Circuits, vol. SC-18, pp. 652-664, Dec. 1983.

8.22] P.W. Li, M.J. Chin, P.R. Gray, and R. Castello, "A ratio-independent algorithmic analog-to-digital conversion technique," IEEE J. Solid-State Circuits, vol. SC-19, pp. 828-836, Dec. 1984.

8.23] B.S. Song and M.F. Tompsett, "A 12b 1 MHz capacitor error averaging pipelined A/D converter," in ISSCC Dig. Tech. Papers, pp. 226-227, San Fransisco, CA, Feb. 1988.

8.24] B. Gilbert, "Translinear Circuits: a proposed classification," Electronic Letters, vol. 11, Jan. 1975.

8.25] M. Soyuer, High-frequency monolithic phase-locked loops, Ph. D. Dissertation, University of California, Feb. 1988.

8.26] K.M. Ware, H.-S. Lee, and C.G. Sodini, "A 200 MHz CMOS PLL," in ISSCC Digest Tech. Papers, New York, N.Y., Feb. 1989.

8.27] M. Kubo, I. Masuda, K. Miyata, and K. Ogiue, "Perspective on BiCMOS VLSI's," IEEE J. Solid State Circuits, vol. SC-23, pp 5-11, Feb. 1988.

8.28] N. Tamba, S. Miyaoka, M. Odaka, K. Ogiue, K. Yamada, T. Ikeda, M. Hirao, H.

Higuchi, H. Uchida, "An 8ns 256K BiCMOS SRAM," in ISSCC Dig. Tech. Papers, pp. 184-185, San Fransisco, CA, Feb. 1988.

8.29] H. V. Tran, D.B. Scott, P.K. Fung, R.H. Havemann, R.E. Eklund, T.E. Ham, R.A. Haken, A. Shah, "An 8ns battery back-up submicron BiCMOS 256K ECL SRAM," in ISSCC Dig. Tech. Papers, pp. 188-189, San Fransisco, CA, Feb. 1988.

Glossary: Symbol Definitions

Symbol	Description	Units
α	Impact ionization coefficient	*
β	BJT current gain = Ic/Ib	*
Δ	Small variation, i.e. ΔVg, ΔId etc	*
ΔI	Δ in MOSFET subthreshold Id for a ΔVd	Dec
ε_o	Permittivity of free space	F/cm
ε_{ox}	Dielectric constant of silicon dioxide	*
ε_{si}	Dielectric constant of silicon	*
τ	Time constant	sec
μ	Mobility	cm^2/Vs
μ_n	Electron mobility	cm^2/Vs
μ_p	Hole mobility	cm^2/Vs
A_C	BJT collector area	cm^2
A_E	BJT emitter area	cm^2
At	Temperature acceleration factor	*
Aef	Electric field acceleration factor	*
BV_{cbo}	Collector-base breakdown voltage|emitter open	V
BV_{ceo}	Collector-emitter breakdown voltage|base open	V
BV_{ebo}	Emitter-base breakdown voltage|collector open	V
BV_{dss}	MOSFET source-drain breakdown with Vgs=0	V
C	Capacitance	F
Ca	Capacitance per unit area	F/cm^2
C_C	NPN collector capacitance	F
C_E	NPN emitter capacitance	F
Cfb	MOS flatband capacitance	F
C_{gd}	MOSFET gate-drain capacitance	F
Cj	Junction capacitance	F
C_{ov}	MOSFET overlap capacitance	F/cm
C_{ox}	MOS oxide capacitance	F
Cp	Capacitance per unit periphery	F/cm
Cs	NPN Collector-substrate capacitance	F
D	Impurity diffusivity	cm^2/sec
E	Electric Field	V/cm
f_T	BJT frequency at which $\beta=1$	Hz
Gds	MOSFET source/drain conductance	$1/\Omega$
Gm	Transconductance	$1/\Omega$
h_{fe}	BJT large signal transistor gain	*
I	Current	A
I_B	BJT base current	A
I_C	BJT collector current	A
I_D	MOSFET drain current	A
I_E	BJT emitter current	A
$I_D sat$	MOSFET saturation current (Vd=Vg=5V)	A

I_G	MOS gate current	A
I_S	BJT saturation current	A
Isub	Substrate current in MOSFETs	A
Ipp	Latchup current	A
I_T	Latchup trigger current	A
I_H	Latchup holding current	A
Ik	NPN Collector Knee Current	A
j_{px}	hole current density	A/cm^2
J_c	BJT critical collector current density for high level	A/cm^2
k	Boltzmann's constant	eV/°K
L	MOSFET channel length	μm
L_{eff}	MOSFET effective channel length	μm
Lf	Effective funneling length	μm
M	Multiplication factor	*
n	BJT ideality factor	*
Na	Substrate (or acceptor) doping concentration	cm^{-3}
Nc	NPN collector doping concentration	cm^{-3}
Nd	Donor doping concentration	cm^{-3}
q	Elementary charge	C
Q	Charge per unit area	C/cm^2
Q_b	NPN base Gummel Number	$1/cm^2$
Q_{bd}	Charge-to-breakdown (in MOS capacitors)	C/cm^2
Q_c	NPN Collector Gummel Number	$1/cm^2$
Q_f	Fixed oxide charge density	C/cm^2
Q_{it}	Interface trap charge density	C/cm^2
Q_m	Mobile ion charge density	C/cm^2
Q_{ot}	Oxide trap charge density	C/cm^2
Qr	Reverse bias stress charge (in NPNs)	C
R	Resistance	Ohm=Ω
R_B	NPN base resistance	Ω
R_{Bext}	NPN extrinsic base resistance	Ω
Rc	Contact resistance	Ω
R_C	BJT collector resistance	Ω
Rch	MOSFET channel resistance	Ω
R_E	BJT emitter resistance	Ω
Rlin	MOSFET linear region resistance	Ω
R_S	MOSFET source/drain resistance	Ω
S_o	Surface recomination velocity	cm/sec
S	Subthreshold swing in MOSFETs	mV/Dec
t	Time	sec
T	Temperature	°C
T_b	Base transit time	sec
T_{ox}	Oxide thickness	cm
Va	BJT Early Voltage	V
V_{BE}	BJT base-emitter voltage	V

V_{CB}	BJT collector-base voltage	V
v_s	Scattering limited velocity	cm/sec
V_D	MOSFET drain voltage	V
V_{Dsat}	MOSFET saturation voltage	V
Vee	Supply voltage	V
V_G	MOSFET gate voltage	V
Vpt	Punch through voltage	V
V_S	MOSFET source voltage	V
V_T	MOSFET threshold voltage	V
W	MOSFET channel width	cm
W_b	BJT basewidth	cm
W_c	BJT effective epi thickness	cm
W_{eff}	MOSFET effective channel width	cm
X_j	Junction depth	cm

About the Authors

Chapter 1. Introduction

Antonio R. Alvarez was born in Havana, Cuba in 1956. He received a BEE and an MSEE from the Georgia Institute of Technology in 1978 and 1979 respectively. He joined Motorola as a staff engineer in 1979 working initially in linear technology development. This included high voltage Bipolar-DMOS I.C.s, Power BIMOS, design rules, yield modeling, and process consolidation. In 1982 he joined Motorola's Bipolar Technology Center specializing in digital bipolar and BiCMOS technologies. Eventually he became responsible for all aspects of BiCMOS Gate Array and Memory technology development. In 1987 he joined Cypress Semiconductor as program manager for BiCMOS technology. When Cypress formed Aspen Semiconductor in 1988 as a new venture for advanced bipolar and BiCMOS products, Mr Alvarez assumed responsibility for Aspens's BiCMOS technology. He is currently Director of Technology and Memory Design at Aspen. Mr. Alvarez has developed and taught courses in applied statistics, basic process and device technology, and BiCMOS. He has published over 20 papers on low and high voltage BiCMOS, application of statistical techniques, MOSFET physics, I.C. material and processing technologies, and has several patents pending. Mr. Alvarez is a member of the IEEE and has served as a subcommittee member for both the International Electron Device Meeting and Symposium on VLSI Technology.

Chapter 2. Device Design

Jim Teplik received the BSEE and MSEE from the University of Cincinnati in 1979 and 1981 respectively. In 1981 he joined Motorola as a Device Engineer for fast static RAMs. In 1985 he transferred to Motorola's Bipolar Technology Center. As a Senior Staff Engineer, his main effort has been the application of statistical methods coupled with process and device simulation tools to develop and optimize BiCMOS technologies. He has coauthored several papers in the area of BiCMOS technology and the application of response surface design. He is a member of Tau Beta Pi and the IEEE.

Chapter 3. BiCMOS Process Technology

Roger A. Haken was born in Purley, Surrey, England in 1950. He received the Higher National Diploma in electrical and electronic engineering from Southampton College of Technology in 1971 and MS and PhD degrees in electronics from the University of Southampton in 1972 and 1975 respectively. From 1975 to 1978 he was with GEC Hirst Research Centre working on design of buried-channel MOSFETs and CCDs, and multilevel polysilicon technology. In 1978 he joined the Central Research Laboratories of Texas Instruments where he was initially involved in the design and fabrication of NMOS and CCD circuits for analog signal processing. In 1980 he became responsible for the development of CMOS technology and circuit design for a SONOS non-volatile RAM. In 1982, he became manager of the VHSIC 1.2μm NMOS SRAM technology in the

Semiconductor Process and Design Center. Since 1984 he has been research manager of TI's 1μm/subμm CMOS technology development for VLSI logic. In 1986 Dr. Haken was elected a TI Fellow and in 1987 assumed the additional responsibility of managing TI's submicron BiCMOS technology. Dr. Haken is a Sr. Member of the IEEE, and past Publications Chairman and subcommittee member of the International Electron Device Meeting. He holds 15 patents and is author or co-author of over 30 technical publications.

Robert H. Haveman received a BA and MS in electrical engineering from Rice University in 1967 and 1970 respectively. He obtained a PhD in electrical engineering in 1974 from the University of Colorado at Boulder. He performed postdoctoral research on thin film optical waveguides at the University of Queensland in Brisbane, Australia. In 1976 he became a National Research Council Postdoctoral Fellow a the National Bureau of Standards in Boulder, Colorado, where his research focused on high speed optical detectors and superconducting I.C.s using Josephson junctions. Since joining Texas Instruments in 1978 he has concentrated on process integration for submicron MOS and bipolar devices. A Senior Member of the Technical Staff, his current responsibilities include BiCMOS process development and implementation. Since 1985 he has also served as an Adjunct Professor of Electrical Engineering at the Southern Methodist University. He holds nine issued U.S. patents and has authored or co-authored 18 publications.

Robert H. Eklund received the BS degree in chemistry from Davidson College in 1974 and the PhD degree in chemistry from the University of Texas at Austin in 1979. After graduating, he joined Texas Instruments where he is currently a Senior Member of the Technical Staff working in the area of process development for submicron Bipolar/BiCMOS technologies.

Louis N. Hutter was born in Covington, Kentucky in 1954. He received the BS degrees in mathematics and physics from Northern Kentucky University in 1976 and the MS degree in electrical engineering from the M.I.T. in 1978. In 1978 he joined Texas Instruments as a process development engineer in the Linear Products department where he was involved in the development of CMOS and high voltage merged technologies. He is presently an engineering manager responsible for advanced CMOS and BiCMOS process development directed towards low and medium voltage analog applications. Elected a Senior Member of the Technical Staff in 1983, Mr. Hutter has authored several articles and has a number of patents pending. He is a member of the IEEE and Sigma Xi.

Chapter 4. Process Reliability
Rajeeva Lahri received the M. Technology in Materials Science from the Indian Institute of Technology and PhD in Electrical Engineering from the State University of New York at Buffalo. He joined Hewlett-Packard's CMOS process development group in 1982 as a Member of the Technical Staff. There he was

involved with the development of various generations of CMOS processes. Dr. Lahri joined National (Fairchild) Semiconductor in 1986. As Engineering Process Manager at NSC Puyallup, Washington, Mr Lahri is responsible for the reliability of ECL and BiCMOS processes. He has been a contributing author to over 15 publications. He is a member of the IEEE and industry mentor for SRC's VLSI reliability research efforts.

Satyapriy Joshi received the M.Sc. degree in Applied Physics from the University of Indore, India and the M.Tech. degree in Materials Science from the Indian Institute of Technology. Mr. Joshi joined the National (Fairchild) Semiconductor Corp., Puyallup, Washington as a Sr. Process Engineer in 1987, where he has been engaged in the characterization, modeling and reliability of advanced BiCMOS and Bipolar processes. He has over 24 publications articles in the area of optoelectronics, compound semiconductors, epitaxial dielectrics and high performance Bipolar, Schottky and BiCMOS technologies. Mr. Joshi is a member of Tau Beta Pi and the IEEE.

Bami Bastani received both MSEE and PhD degrees in solid state electronics from Ohio State University. He joined Intel Corporation's Technology Development group in 1980 and worked on the 64K DRAM project as a Senior Device Physicist. Later he became a Project Manager for 1 Megabit DRAM and Advanced Microprocessor technologies. Dr. Bastani joined National (Fairchild) Semiconductor since 1985. He is the Director of Technology Development at NSC Puyallup, Washington. He is currently concentrating on ECL and BiCMOS technologies for high performance and high density memory and logic applications. He is a member of the IEEE and on the Technical committees for the Technology Symposium on VLSI and the Device Research Conference.

Chapter 5. BiCMOS Digital Design
Kevin Deierling is currently with the programmable logic group of Dallas Semiconductor designing non-volatile lithium backed integrated circuits. He received his BA in physics from the University of California at Berkeley in 1985. He was previously employed with Motorola's ASIC division designing high speed ECL, TTL, and CMOS compatible BiCMOS gate arrays. Prior to that he was employed with U.C. Berkeley's Department of Electrical Engineering working on projects such as voice recognition and automated interfaces.

Chapter 6. Standard Memories
Hiep van Tran was born in Viet Nam in January 1949. From 1978 to 1985 he was with Mostek working as an I.C. circuit designer. He joined Texas Instruments in 1985 and is currently a Member of the Technical Staff in the VLSI Design Laboratory. Mr. Tran has been awarded ten U.S. patents.

Pak Kuen Fung received the BS with honors and the MS degrees in electrical engineering form the University of Arkansas in 1984 and 1986 respectively. While at the University of Arkansas Mr. Fung was supported by a TI Fellowship. He joined Texas Instruments in 1983 part time and full time in 1986 at the VLSI Design Laboratory. He is currently involved in SRAM circuit design using BiCMOS technology. Since joining TI, he has helped design two 16K BiCMOS SRAMs and a 256K BiCMOS SRAM. Mr. Fung is a member of Tau Beta Pi and Eta Kappa Nu.

David B. Scott received the BSc degree in physics in 1974 and the MASc and PhD degrees in electrical engineering in 1976 and 1980 respectively all from the University of Waterloo. In 1979 he joined Texas Instruments as a Member of the Technical Staff. Currently he is a Senior Member of the Technical Staff in the VLSI Design Laboratory working on BiCMOS modeling and design. He has published papers on short-channel effects in MOSFETs, charge-coupled and bucket-brigade devices, bipolar & CMOS technology, silicide contact resistance, hot carrier effects and electrostatic discharge protection. His current research interest is in BiCMOS design techniques.

Ashwin Shah was born in Bombay, India in 1950. He received the B. Technology degree in electrical engineering from the Indian Institute of Technology in 1972 and the MS degree in electrical engineering from the Illinois Institute of Technology in 1975. From 1974 to 1975 he was involved in DRAM design at Mostek. He joined Texas Instruments in 1975 and since then has been involved in the design and development of CCD and other high density MOS memories. He has been involved in various high density DRAM designs and managed the design activities of an experimental 4Mbit DRAM. His current research interests are in the area of BiCMOS applications for memories and ASICs, presently managing the BiCMOS design branch of the VLSI Design Lab in the Semiconductor Process and Design Center at TI. Mr. Shah was elected a TI Fellow in 1988. He is a senior member of the IEEE, has been a member of the ISSCC program committee since 1986, and is a member of the Symposium on VLSI Circuits.

Chapter 7. Specialty Memories

Charles Hochstedler received an BS in electrical engineering technology with honors from Purdue University in 1976. He then joined Harris Semiconductor as an Applications Engineer, where he became interested in new product definition. Since then he has held engineering and technical marketing positions at Intel and Mostek. Responsible for defining new memory products at Fairchild when the company was acquired by National Semiconductor, currently he is the Product Planning Manager for the high performance BiCMOS memory product line at National. Mr. Hochstedler has published several papers in technical conferences and articles in trade publications. He is a member and task group chairman in the JEDEC committees chartered with memory pinout standardization.

Chapter 8. Analog Design

Hae-Seung Lee was born in Seoul, South Korea in 1955. He received the BS and MS degrees in electrical engineering from the Seoul National University in 1978 and 1980 respectively. He received the PhD degree in electrical engineering from the University of California at Berkeley in 1984, where he worked on self-calibration techniques for A/D converter. In 1980, he was a Member of the Technical Staff in the Department of Mechanical Engineering at the Korean Institute of Science and Technology where he was involved in the development of alternative energy sources. Since 1984 he has been with the Department of Electrical Engineering and Computer Science at the Massachusetts Institute of Technology, where he is now an Associate Professor. Since 1985, he has served as a consultant to Analog Devices and MIT Lincoln Laboratories. His research interests are in the areas of integrated circuits, devices, fabrication technologies, and solid-state sensors. Prof. Lee is the recipient of a 1988 Presidential Young Investigator's Award.

Index